SpringerBriefs in Plant Science

More information about this series at http://www.springer.com/series/10080

Girdhar K. Pandey • Manisha Sharma
Amita Pandey • Thiruvenkadam Shanmugam

GTPases

Versatile Regulators of Signal Transduction in Plants

 Springer

Girdhar K. Pandey
Department of Plant Molecular Biology
Delhi University South Campus
Dhaula Kuan, New Delhi, India

Amita Pandey
Department of Plant Molecular Biology
Delhi University South Campus
Dhaula Kuan, New Delhi, India

Manisha Sharma
Department of Plant Molecular Biology
Delhi University South Campus
Dhaula Kuan, New Delhi, India

Thiruvenkadam Shanmugam
Division of Biosciences and Bioinformatics
Myongji University
Kyunggi-do, Republic of South Korea

ISSN 2192-1229   ISSN 2192-1210 (electronic)
ISBN 978-3-319-11610-5   ISBN 978-3-319-11611-2 (eBook)
DOI 10.1007/978-3-319-11611-2
Springer Cham Heidelberg New York Dordrecht London

Library of Congress Control Number: 2014951895

Printed on acid-free paper

Springer is part of Springer Science+Business Media (www.springer.com)

Preface

In the signaling pathways, the activation or inactivation of the proteins is determined by several regulatory components. One of the major regulatory controls is phosphorylation–dephosphorylation cascade mediated by kinases and phosphatases; besides this, G proteins including heterotrimeric and small GTPases also act as essential regulatory switch in the modulation of these signaling pathways. Rho family of GTP-binding proteins (GTPases) acts as binary molecular switches that mediate large number of intracellular signals in eukaryotes. They acquire an activated conformation when bound to GTP (guanosine triphosphate) and are inactivated by hydrolysis of GTP to GDP (guanosine diphosphate). In recent years, a wealth of information has been generated for understanding Rho protein functions in plants. Accordingly, GTPases are instrumental in relaying signals ranging from actin and microtubule arrangement, cell cycle progression, vesicle trafficking, cell morphology, and root hair elongation in plants.

Chapter 1 provides an overview of small GTPases in eukaryotes. The small GTPase superfamily has evolved enormously in metazoan lineage and was classified into five subfamilies (Ras, Rab, Rho, Ran, and Arf) based on their distinct functions in the cell. Three different regulatory proteins (GEFs, GAPs, and GDIs) control the nucleotide state of Rho proteins. GEFs (guanine nucleotide exchange factors) are the activation factors that catalyze the exchange of GDP for GTP. On the other hand, GAPs (GTPase activating proteins) cause Rho proteins inactivation by inducing their intrinsic GTP hydrolysis activity. Finally, GDIs (guanine nucleotide dissociation inhibitors) show specific affinity for inactivated GTPases and prevent them from further activation. Chapter 2 conveys an overview of Rho GTPases in plants and also discusses their known functional role and cross talk in myriad of signaling pathways.

Among the six Ras superfamily GTPases classified in animals, five have been identified in plants, whereas Ras subfamily of GTPases is altogether absent in plants. Additionally, Cdc42 and Rho subfamilies are absent in plants, but instead they possess a novel group of Rac-like signaling molecules, also known as ROP GTPases. More than 90 ROP proteins have been identified in *Arabidopsis*, and with an extensive database search, we could identify 85 ROPs in *Oryza sativa*. Chapter 3 covers the identification and classification of ROP GTPases in plants.

The evolution of functionally distinct Rac-like GTPases in plants, and, furthermore, due to several gene duplication events, bifurcations of these into distinct subfamilies in both monocots and dicots have generated interest towards their phylogenic evolution. The detailed comparative phyletic and correlative analyses between plants and animals as well as their domain organization have been included in Chapter 4.

During the past several years, remarkable progress has been made towards elucidation of functions that are mediated by Rho proteins in plants. It is not surprising that the immense cellular functions of ROP proteins in plants encompass developmental and stress responses as well. Chapter 5 consists of the expression analysis of identified Rho GTPases in *Arabidopsis* and rice under stress, development, and phytohormone treatment that would be beneficial for gleaning out their specialized and overlapping functional role. Since then several studies have recognized numerous signaling pathways that are controlled by Rho proteins. Chapter 6 lists some of the extensively studied and essential roles of ROPs in plants.

A requisite for the suitable subcellular localization of Rho family GTPases is their posttranslational lipid modification by hydrophobic side groups. The prenylation and palmitoylation of the C-terminal CAAX motif is needed as a lipid anchor to facilitate their plasma membrane association. Chapter 7 deciphers the mechanism of posttranslational lipid modification and membrane association of ROP GTPases in plants. The regulatory mechanism of Rho GTPases and their regulator and effector molecules are discussed in Chapter 8.

Meanwhile, researchers have put a concerted effort to develop new methods and techniques to study GTPases and their roles in plants. It was speculated that, since GTPases exist as a multigenic family, they might be functionally redundant and are possibly involved in signaling cross talk. The level of functional intricacy displayed by GTPases creates complications in their structural study. Several new genetic and biochemical approaches have been devised to study their biological functions. Chapter 9 reviews some of the promising prevailing techniques to study GTPases in living cells. Finally, the future prospects including importance of elucidation of regulatory mechanism of ROP proteins to get an insight into their core principles and actions have been discussed in Chapter 10.

Rho GTPases signaling pathways are a model for cell biologists to elucidate signal transduction pathways. We hope this book will prove beneficial to both students and researchers in this field and will enable them to understand the mechanisms and importance of these versatile signaling molecules in plants.

New Delhi, India   Girdhar K. Pandey
New Delhi, India   Manisha Sharma
New Delhi, India   Amita Pandey
Yongin, Kyunggi-do, Republic of South Korea   Thiruvenkadam Shanmugam

Acknowledgement

We are thankful to the University of Delhi, University Grants Commission (UGC), Department of Science and Technology (DST) and Department of Biotechnology (DBT), India, for supporting the research work in GKP's lab.

Contents

Chapter 1
Overview of G Proteins (GTP-Binding Proteins) in Eukaryotes

Introduction

There is a spectrum of small GTP-binding proteins (G proteins), ranging in size from 20 to 40 kDa, present in eukaryotic cells that utilize the binding and hydrolysis of GTP. By virtue of its binding and hydrolysis, the G proteins tentatively behave as molecular switches and this phenomenon is the basis for many ubiquitous regulatory processes in eukaryotes [1]. The high degree of sequential conservation of G proteins among eukaryotes underscores the similarities in functional control of cellular processes. They regulate diverse cellular processes like protein synthesis, early and late secretory pathway, inter- and intracellular signal transduction, cell proliferation, and differentiation [2].

Based on their subunit structure and molecular weight, these can be divided into heterotrimeric G proteins and Ras superfamily of monomeric small GTPases. The Ras superfamily in humans has a catalogue of 150 proteins that are also conserved in *Drosophila*, *C. elegans*, *Dictyostelium*, and plants [3]. The Ras proteins were identified as mutated forms of oncogenes that stimulate proliferation of cultured cells. They were discovered early owing to their high oncogenic potential when transduced into retroviruses like the Harvey and Kirsten sarcoma viruses that possess H-ras and K-ras, respectively. Even some mutated forms affected differentiation of neuronal cells [4]. In the case of yeast, two genes were identified, Ras1 and Ras2, that are critical for viability. More importantly, the Ras mutants of yeast could be complemented by human homologs [5]. Even though Ras oncogenes were the first ones to be identified, the whole superfamily is divided into five subfamilies: Ras, Rho, Rab, Arf/SAR, and Ran. Active Ras proteins can switch on/off, variety of downstream effectors, thereby regulating gene expression networks and cytoplasmic signaling to control cell proliferation, differentiation, and viability. There are a variety of complex pathways regulated by small GTPases owing to the array of posttranslational modifications, differential subcellular localization, and effector/regulators [2].

© The Author(s) 2015
G.K. Pandey et al., *GTPases*, SpringerBriefs in Plant Science,
DOI 10.1007/978-3-319-11611-2_1

The discovery of homologs of Ras proteins (Rho) in yeast, *YPT1* and marine snail Aplysia, *Rho*, leads to the finding that these proteins share about 30 % homology to the Ras family of proteins. The mutant of yeast *YPT1* is defective in the budding process during the life cycle, indicating the possibility of defective cytoskeleton reorganization [6]. In the mammalian Rho proteins, Rac1 was first identified to be required for the activation of NADPH oxidase of phagocytic cells [4]. Many of the mammalian Rho proteins were figured using a C3 exoenzyme of *Clostridium* that can uniquely ADP-ribosylate Rho proteins at specific amino acid of effector region, which could prevent the interaction with its downstream regulators [7]. These proteins were further demonstrated to be involved in stress fiber modification and Ca^{2+} regulation during smooth muscle contraction [4]. Further, their roles have been established in integrating the extracellular signal information into the gene expression circuitry.

The Rab proteins were first described as Ras-related proteins in brain. They form the largest subfamily contributing to Ras superfamily in most eukaryotes. These were identified as conserved chief regulators of intracellular vesicle trafficking in yeast. Along with Rab, SEC4 and Ypt1 proteins in yeast were shown to control the vesicle transport between Golgi and plasma membrane [8]. They promote vesicle formation by facilitating budding from the donor membrane, targeting to the acceptor compartment, and releasing of the vesicle into the receptor compartment. Their function starts with the localization of the Rab proteins in distinct intracellular compartments. Interestingly, this localization is dependent on the level of prenylation of the protein and divergence of C-terminal domain [2]. In yeast, many Rab proteins are reported to be involved in cell viability.

ARF (ADP-ribosylation factors)/SAR1 (secretion-associated RAS-related protein 1) family of proteins is closely related with the Rab proteins in terms of its function especially on vesicle transport. The active form of Arf can interact with vesicle coat proteins to regulate distinct downstream regulators. This interaction promotes sorting of cargos while affecting formation and release of vesicle. While Rab controls any single step in vesicle trafficking, Arf has been shown to regulate multiple stages of vesicular transport. The SAR1 gene of yeast has shown to be involved in ER–Golgi network transport by COPII-mediated pathway affecting its assembly and disassembly [9].

The Ran (Ras-like nuclear protein) family of small GTPase is perhaps the smallest subgroup and its function is envisaged to regulate nucleocytoplasmic transport. The evidence is based on the mutants' inability to import a reporter construct containing nuclear localization signal of a simian virus 40T antigen. They are structurally similar to Rab family of proteins but have distinct features like spatial gradient regulation of active GTP-bound form meaning they are asymmetrically distributed between nucleus and cytoplasm [10]. The Ran protein facilitates nuclear import and export controlled by upstream activators and downstream effectors. This is usually achieved by its interaction with importin and exportin to promote cargo import and export, respectively. Ran proteins lack sites for posttranslational modifications unlike other small GTPases and hence do not require lipids for membrane binding or for its activation [11].

Small G Protein Structure and Domain features

Small GTPases are monomeric molecules that can form stable complex by binding to GTP or GDP, while they are poor catalysts on their own and require GTPase activating protein (GAP) for their inactivation. The active or inactive state of these proteins is defined by their ability to bind GTP or GDP at any given time, where guanine nucleotide binding causes distinct conformational changes in the protein structure. The available three-dimensional structures of small G proteins have shed light on the protein regulation and activity. Approximately, 20 kDa size of conserved domain (G domain) is responsible for binding and hydrolysis of guanine nucleotides. The domain is built of five alpha-helices (denoted $\alpha 1-\alpha 5$), six beta-strands (denoted $\beta 1-\beta 6$), and five hydrophobic loops (denoted G1–G5). Perhaps the contrarian of G domain is that the loops (G1–G5) are more conserved than helices and sheets. The conservation pattern of G domain beginning at the N-terminus is: G1, GXXXXGKS/T; G2, T; G3, DXXGQ/H/T; G4, T/NKXD; G5, C/SAK/L/T [12, 13]. The comparison of loop structures between GTP-bound and GDP-bound forms revealed two distinctive functional regions: Switch I and Switch II flanking the gamma-phosphate of the guanine nucleotide [14]. The loop connecting $\beta 1$ strand and $\alpha 1$ helix is G1 loop (alternatively, P loop) responsible for binding to α- and β-phosphate groups. Another distant loop, G3 loop, provides binding elements to Mg^{2+} and γ-phosphate groups. In fact, these two G1 and G3 loops hold structural similarity to Walker A and Walker B boxes of erstwhile nucleotide-binding motifs that are not related to small G proteins. The G4 and G5 loops account for the specificity of guanine residue. The Lys and Asp residues of G4 conserved loop directly bind with the nucleotide while a part of G5 loop is held for guanine specificity [1]. Comparison of several G protein structures would reveal that G domain forms the basal structure for all these functional similarities while variations could be accounted on this canonical structure. With the structural availability of GTP-bound and GDP-bound G domains, the prerequisite for molecular switch has been defined. The dynamics of structural changes from Switch I to Switch II differs significantly between GTP- and GDP-bound forms in NMR (nuclear magnetic resonance), EPR (electron paramagnetic resonance), and FTIR (Fourier-transform infrared spectroscopic investigation) spectroscopy [15]. The minimal G domain catalytic apparatus maintains the ability to hydrolyze GTP, and the released energy could be utilized for the conformational change in the effector regions that are suffice for the cycling of alternate forms. Among five loops, three loops, G1, G2, and G4, are flexible enough to allow localized polysterism critical for the functioning of G domain apparatus [15]. The dissimilarities among these small GTPase superfamily members mainly border on variations in nucleotide-binding region including extra α-helix in N-terminal region (like in Rho proteins), antiparallel β-sheet in switches I and II (like in Arf proteins), and ways to coordinate magnesium ion (like in Arf proteins) [1].

Biochemical Regulation of Small GTPases

Their readily available protein structures provide the clue that they exist in at least two forms: GTP-bound active and GDP-bound inactive GTPases. The signaling pathways regulated by GTPases emphasize the significance of external signals in converting their inactive form to activated form. There are upstream regulators like cell membrane receptor protein complexes that favor the dissociation of GDP to become GTP-bound form. This GTP-bound form can cause conformational changes to the protein, essentially triggering the downstream effector regions. The intrinsic GTPase activity can revert this conformational change by displacing GTP with GDP. This activation and inactivation brings a cycle complete for the small GTPase. To regulate this cycle positively and negatively, there are many dedicated proteins in eukaryotes readily available favoring GTP-bound or GDP-bound form [16, 17]. The proteins that facilitate the displacement of GDP with GTP were named guanine exchange factors (GEF) that are chiefly positive regulators for small GTPase activity. In this reaction, GEFs bind with Ras-GDP to release GDP while forming a binary complex with the small GTPase that can facilitate GTP to bind small GTPase. A lot of cooperativity has been reported with GEFs where one GEF can transduce signal to Ras as well as Rho proteins playing a critical role in cross-talk signaling mechanisms. At the same time, a signal from a single receptor could be amplified by two or more GEFs. This type of versatility exists to accommodate signal divergence or convergence.

Along with positive regulation, there are another two functionally distinct groups of proteins existing to negatively regulate the small GTPase activity by dissociating the GTP from the active form with GDP. The negative regulation is mainly the resultant of GTPase dissociation inhibitors (GDIs) and GAPs. As the name suggests, GAP proteins promote the intrinsic GTPase activity, which promotes the rate of GTP hydrolysis activity converting it into GDP-bound inactive form. GDIs act as negative regulators by sequestering the available small GTPase proteins in cell membrane by binding with them in the cytosol. The mode of sequestration keeps the small GTPase in inactive GDP-bound form. This model of secluding the available GTPase from the membrane thereby creates a differential GTPase pool between cytosol and membrane and is physiologically significant than favoring the actual biochemical GTP dissociation [1, 4].

Guanine Exchange Factors

GEF proteins accelerate the dissociation of GDP bound to the small G protein and therefore facilitate its conversion to active GTP-bound state. GEFs are multiple domain-containing proteins that are critical for its protein–protein and protein–lipid interactions. GEFs have a catalytic DH domain arranged in tandem with a PH domain. PH domain is a common feature of signaling molecules where an interaction with this domain could target the complex to membrane. It is also suggested that PH domain could fulfill a catalytic role assisting DH domain for GEF activity [1, 18].

GTPase Activating Proteins

GAP proteins, dominantly made up of α-helices, play a role in surging the rate of GTP hydrolase activity up to 4–8 orders of magnitude. Switch I and Switch II of G proteins contribute to GAP proteins' interaction. This surge in hydrolysis is possible due to an Arg residue supplied by GAP protein α-helices. BH domains of GAP proteins' C-terminal region interact with GTPase proteins. Additionally, SH2, SH3, and PH domains contribute to GAP function. Catalytic residues are positioned in GAPs to accelerate GTP hydrolysis [4, 19] .

GTPase Dissociation Inhibitors

GDI proteins stand out from other GTPase regulators that they recognize posttranslational modifications on small GTPase. So much so that small GTPases occur as GDI-bound form in cytosol. It recognizes the geranylgeranyl moiety on G proteins for binding. GDI function in general does not share structural similarities: like in RabGDI and RhoGDI. One isoform of GDI, α-isoform, is made up of two domains: smaller α-helical domain and a larger β-domain. The larger domain contains GCD domain responsible for interaction with GTPases and a conserved region for binding with posttranslational modifying enzymes. The smaller domain structure resembles monooxygenases [1, 4, 20].

Localization and Posttranslational Modifications

A few members of small GTPase display tissue-specific expression: Rab17 of humans is expressed in epithelial cells while Rab3A is expressed in secretion pathway cells such as neurons, neuroexocrine cells, and neuroendocrine cells. Most G proteins are present either in cytosol or nucleus. RanGTPase, however, is distributed between cytosol and nucleus. Mammalian Ras proteins are present on the lower membranes facing the cytosol. Posttranslational lipid modifications on small GTPases play a significant role in targeting these proteins to cell membranes. Majority of Ras and Rho proteins end with a tetrapeptide on their C-terminal region comprising a conserved code CAAX (C, Cys; A, aliphatic residues; X, any residues). The amino acid sequence immediately upstream of cysteine residue undergoes lipid modification where the conserved tetrapeptide is recognized by modifying enzymes like geranylgeranyltransferase I and farnesyl transferase [2]. Other Rab family members possess other C-termini residues such as CC, CCX, CXC, CCXX, and CCXXX that are recognized by geranylgeranyltransferase II. Members of Arf family are modified with myristate group on its N-termini region with all these lipid modifications becoming critical for biological activity of small GTPases. However, Ran proteins are not subjected to lipid modifications since they are not membrane-bound proteins [2, 21].

References

1. Paduch M, Jelen F, Otlewski J. Structure of small G proteins and their regulators. Acta Biochim Pol. 2001;48(4):829–50.
2. Wennerberg K, Rossman KL, Der CJ. The Ras superfamily at a glance. J Cell Sci. 2005;118 (Pt 5):843–6.
3. Colicelli J. Human RAS, superfamily proteins and related GTPases. Sci STKE. 2004; 2004(250):RE13.
4. Takai Y, Sasaki T, Matozaki T. Small GTP-binding proteins. Physiol Rev. 2001;81(1): 153–208.
5. Powers S, Kataoka T, Fasano O, Goldfarb M, Strathern J, Broach J, et al. Genes in S. cerevisiae encoding proteins with domains homologous to the mammalian ras proteins. Cell. 1984;36(3):607–12.
6. Segev N, Mulholland J, Botstein D. The yeast GTP-binding YPT1 protein and a mammalian counterpart are associated with the secretion machinery. Cell. 1988;52(6):915–24.
7. Chardin P, Boquet P, Madaule P, Popoff MR, Rubin EJ, Gill DM. The mammalian G protein rhoC is ADP-ribosylated by Clostridium botulinum exoenzyme C3 and affects actin microfila-ments in Vero cells. EMBO J. 1989;8(4):1087–92.
8. Du LL, Collins RN, Novick PJ. Identification of a Sec4p GTPase-activating protein (GAP) as a novel member of a Rab GAP family. J Biol Chem. 1998;273(6):3253–6.
9. Schmid SL, Damke H. Coated vesicles: a diversity of form and function. FASEB J. 1995;9(14):1445–53.
10. Moore MS. Ran and nuclear transport. J Biol Chem. 1998;273(36):22857–60.
11. Ullman KS, Powers MA, Forbes DJ. Nuclear export receptors: from importin to exportin. Cell. 1997;90(6):967–70.
12. Bourne HR, Sanders DA, McCormick F. The GTPase superfamily: conserved structure and molecular mechanism. Nature. 1991;349(6305):117–27.
13. Dever TE, Glynias MJ, Merrick WC. GTP-binding domain: three consensus sequence elements with distinct spacing. Proc Natl Acad Sci U S A. 1987;84(7):1814–8.
14. Wei Y, Zhang Y, Derewenda U, Liu X, Minor W, Nakamoto RK, et al. Crystal structure of RhoA-GDP and its functional implications. Nat Struct Biol. 1997;4(9):699–703.
15. Ihara K, Muraguchi S, Kato M, Shimizu T, Shirakawa M, Kuroda S, et al. Crystal structure of human RhoA in a dominantly active form complexed with a GTP analogue. J Biol Chem. 1998;273(16):9656–66.
16. Tong L, de Vos AM, Brunger A, Yamaizumi Z, Nishimura S, et al. Molecular switch for signal transduction: structural differences between active and inactive forms of protooncogenic ras proteins. Science. 1990;247(4945):939–45.
17. Vetter IR, Wittinghofer A. The guanine nucleotide-binding switch in three dimensions. Science. 2001;294(5545):1299–304.
18. Yu H, Schreiber SL. Structure of guanine-nucleotide-exchange factor human Mss4 and identi-fication of its Rab-interacting surface. Nature. 1995;376(6543):788–91.

19. Tesmer JJ, Berman DM, Gilman AG, Sprang SR. Structure of RGS4 bound to AlF4-activated G(i alpha1): stabilization of the transition state for GTP hydrolysis. Cell. 1997;89(2):251–61.
20. Lian LY, Barsukov I, Golovanov AP, Hawkins DI, Badii R, Sze KH, et al. Mapping the binding site for the GTP-binding protein Rac-1 on its inhibitor RhoGDI-1. Structure. 2000;8(1): 47–55.
21. Seabra MC, Wasmeier C. Controlling the location and activation of Rab GTPases. Curr Opin Cell Biol. 2004;16(4):451–7.

Chapter 2
Overview of Small GTPase Signaling Proteins in Plants

Introduction

During the past few years, studies on plant RHO-type (ROP) GTPases have generated new insights into their role in diverse processes such as cytoskeletal organization, polar growth, and development to stress and hormonal responses. Studies have shown that plants have evolved specific regulators and effector molecules. ROP GTPases possess the ability to interact with these multiple regulator and effector molecules that ultimately determine their signaling specificity. Recently, genome wide studies in plants have shown that the *Arabidopsis* genome encodes 93 and rice has nearly 85 small GTPase homologs. We have been able to identify four new homologs in the rice genome. Here, we are focusing on the complete phylogenetic, domain, structural, and expression analyses during stress and various developmental processes of small GTPases in plants. The comparison of gene expression patterns of the individual members of the GTPase family may help to reveal potential plant-specific signaling mechanisms and their relevance. Also, we are summarizing the role of currently known ROP GTPases and their interacting proteins with brief description, simultaneously, comparing their expression pattern based on microarray data. Overall, we will be discussing the functional genomics perspective of plant Rho-like GTPases and their role in regulating several physiological processes such as stress, hormone, pollen tube, root hair growth, and other developmental responses.

Historical Aspects

The discovery of G proteins as the pivotal signaling molecule in 1980 has turned out to be a major breakthrough in this area of research [1]. Since their discovery, contrary to other subgroups ROP/RACs have been subjected to intense research due to their multifunctional role.

© The Author(s) 2015
G.K. Pandey et al., *GTPases*, SpringerBriefs in Plant Science,
DOI 10.1007/978-3-319-11611-2_2

Fig 2.1 Schematic diagram
of the ROP GTPase signaling
in plant cell. Different
extracellular stimuli are
perceived by the receptor
(putative receptor-like
kinases) to activate the
membrane-bound ROPGEFs.
GEFs are the activating
factors of ROPs by catalyzing
GDP to GTP exchange

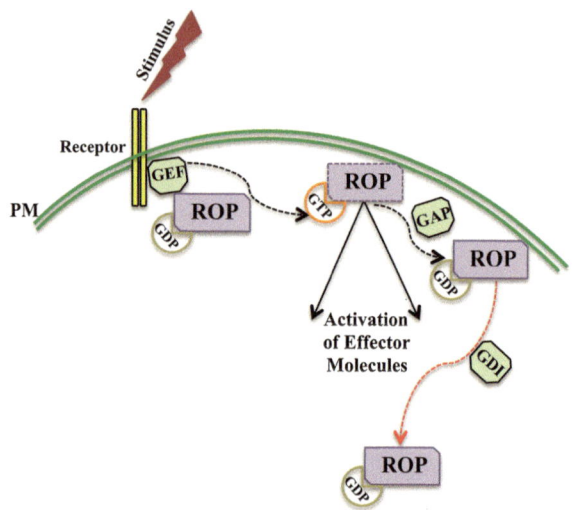

GTPases act as molecular switches where binding with GTP causes an "active" state transition and hydrolysis of GTP to GDP renders them back to an "inactive" state. GTPases exist ubiquitously in eukaryotes constituting a superfamily with five subfamilies Ras, Rho, Rab, Sar1/Arf, and Ran and are among the largest known families of signaling proteins [2]. G proteins are divided into two types, heterotrimeric and monomeric GTPases based on the composition of subunits as well as on their relative molecular mass [3–5]. The heterotrimeric GTPases have been implicated in diverse cellular responses in animals but their functions are not as widespread as in plants. Intriguingly, attempts to identify small GTPases in plants have failed to find any member of Ras GTPase subfamily; alternatively, they have a unique subfamily of Rho-family GTPases, called ROPs (Rho-related GTPases from plants) [6]. The regulatory function of Rho GTPases has been found to be evolutionary conserved while they also act as the master switches for the transmission of extracellular and intracellular signals in plants [4, 5] (Fig. 2.1).

Similar to animals and yeast, Rho GTPases interact with several upstream regulator and effector molecules. Perhaps, the ability to interact with multiple interactors accounts for the functional versatility of small GTPases in plants [7]. Functional conservation of Rho GTPase in plants and animals also extends to the upstream and downstream interactor proteins, yet plants also have evolved unique regulatory GEFs (guanine nucleotide exchange factors) and effector proteins perhaps with novel functions. Emerging roles for ROP/RAC signaling in plants include various developmental responses such as cell morphogenesis, polarized cell growth, endocytosis, and meristem maintenance [8–10]. Moreover, a subset of G proteins is also found to be associated with stress responses such as oxygen deprivation responses and hormonal and defense responses [11–15]. Existence of varying levels of functional divergence in plants indicates a possible gain of new or related function within them [16]. Thus, with the availability of information of structure, sequences, and expression of GTPases in representing members of plant species, it will be interesting to look into the degree of expansion and conservation of these genes in different lineages. A detailed functional genomics account of ROP in plants will unearth the complexity of regulation of various physiological and developmental processes.

Small GTPase Complement in Plantae

Small GTPase superfamily endows plants with the ability to modulate several plant-specific cellular and molecular processes. Even though small GTPases are well conserved across eukaryotic lineage at both sequential and functional levels, sporadic lineage-specific functional variation in some plant species has been observed. In plants, small GTP-binding genes were generally grouped in four main subfamilies: Arf/Sar, Rab, Rop, and Ran [16]. In addition to four main subfamilies, a distinct class of GTPases named as MIRO GTPases has also been recently identified in plants [17]. The physiological functions of many of GTPases have been studied in plants. As assumed, all of the conserved GTPase proteins regulate same general processes in plants as well as in animals [18–20].

The Arf subfamily of GTPases functions as regulators of membrane trafficking and actin remodeling [21, 22]. Many studies based on Arf GEFs and ARF GAPs have elucidated their involvement in protein trafficking from cytosol to the plasma membrane. GNOM has been identified as an Arf GEF essential for targeting PIN1 (PIN-FORMED), an auxin efflux carrier to polar regions [23]. Similarly, ARF GAPs were reported to mediate AUX1 (AUXIN-RESISTANT 1) endosome trafficking to regulate auxin-mediated plant development stimulated by microfilament disruptions [24].

Remarkably, Rab subfamily of small GTPases represents the highest number of GTPase proteins in plants. *Arabidopsis* genome encodes 57 Rab proteins divided into eight distinct clades [25]. Moreover, most of the Rab proteins in these clades have functionally evolved to the extent that each clade could contain distinct proteins [25, 26]. Many functional studies centered on Rab GTPases have confirmed their potential role in endosome organization, post-Golgi targeting to the plasma membrane and vacuoles, and in cytokinesis [27–29]. As seen in tomato, expression of antisense RNA of Rab11 in plants inhibits secretion of an important enzyme suggesting a role in bona fide secretory trafficking pathway.

Ras and Rho act as the signal transducers in animals and in lower eukaryotes [30]. However, no Ras subfamily representative has been identified to be encoded in the plant genome. Rho signaling proteins are functionally diverse and control gene expression, ROS production, cell wall synthesis, vacuolar trafficking, and cell differentiation in eukaryotes [31, 32]. Remarkably, none of the plant Rho GTPases are direct homolog of any of the animal and fungal Rho GTPases. Instead, plants contain a unique subfamily of plant Rho-like GTPases named Rop (Rho-related GTPases from plants).

Possibly, the origin of Rops in plants occurred prior or subsequent to the evolution of angiosperms since lower nonvascular plant such as *Physcomitrella patens* also encodes three ROP genes in its genome [33]. The large number of ROP proteins suggests their implication in numerous pathways as well as potential functional redundancy.

Functional versatility of ROPs required them to interact with various upstream regulators and downstream effectors in both plant and non-plant organisms. As anticipated, few of these interactors are conserved throughout eukaryotes but many of these ROP regulators are specific to plants, coherent to the plant-specific Rho GTPases in plants [7, 34]. The activating RhoGEFs family is divided into two classes in animals. One class also known as Dbl family consists of DH and PH domains [35], which are absent in plants. The second class includes two conserved domains named as dock homology regions 1 and 2 [35]. A dock-like RhoGEF has also been found in *Arabidopsis* called SPIKE1 (SPK1) functioning as activating GEF for ROP proteins [36]. Apart from this, plants have evolved a unique subfamily of RhoGEFs also known as ROPGEFs within them for the activation of ROP proteins [37, 38]. These distinct forms of GEFs, also known as PRONE (plant-specific ROP nucleotide exchanger domain) proteins, bear absolutely no homology with the animal GEFs [37, 38]. In *Arabidopsis*, nearly 14 ROPGEFs have been found to consist a central PRONE domain [38].

In animals, relaying of extracellular signaling responses through plasma membrane to RhoGTPases is facilitated by transmembrane receptor tyrosine kinases (RTKs) [39]. Interestingly, RTK family is not found in plants. However, plants encode a large family of receptor-like kinases (RLKs) with different extracellular and serine-threonine kinase cytoplasmic domain proteins implicated in myriad of signaling responses [40, 41]. Beside, various specific components of signaling responses, plants also constitute several novel ROP interacting partners. For example, in rice, OsRAC1 shows specific interaction with OsCCR1, which is an enzyme involved in lignin biosynthesis [42]. This suggests that plants indeed have evolved many novel ROP regulators, effectors, and signaling pathways, which are yet to be characterized.

Plant-Specific Functions of ROPs

Even though phylogenetic analysis of ROP/RAC proteins in eukaryotes suggests a distinct ROP subfamily in plants, many regulatory components such as RhoGAP, RhoGDI, and dock-type RhoGEF essential for ROP activity were found to be conserved between them. However, few pathways unique to plants mediating signaling response from plasma membrane to ROP were also found to exist. In addition, a RLK and a RhoGEF family were found to exist exclusively in plants. ROP proteins in plants and animals perform similar functions such as polarity establishment, ROS production, and cell morphogenesis irrespective of their mechanisms. At the same time, plant ROPs are specialized for functions unique to plants such as lignin biosynthesis.

Although functional characterization has been carried out for some of the ROP proteins, there is still very limited knowledge about other members. Among the effector proteins RICs belong to the most interesting group mediating actin dynamics and calcium levels in the cytosol. More studies are needed to decipher the detailed mechanism by which they regulate cross talk between different signaling pathways. Additionally, in-depth analysis of processes such as hormonal responses, stress, and defense responses may aid in identifying unknown effectors of ROPs in plants.

References

1. Northup JK, Sternweis PC, Smigel MD, Schleifer LS, Ross EM, Gilman AG. Purification of the regulatory component of adenylate cyclase. Proc Natl Acad Sci U S A. 1980;77(11): 6516–20.
2. Nagano F, Sasaki T, Takai Y. Purification and properties of Rab3 GTPase-activating protein. Methods Enzymol. 2001;329:67–75.
3. Assmann SM. Heterotrimeric and unconventional GTP binding proteins in plant cell signaling. Plant Cell. 2002;14(Suppl):S355–73.
4. Bourne HR, Sanders DA, McCormick F. The GTPase superfamily: a conserved switch for diverse cell functions. Nature. 1990;348(6297):125–32.
5. Bourne HR, Sanders DA, McCormick F. The GTPase superfamily: conserved structure and molecular mechanism. Nature. 1991;349(6305):117–27.
6. Meier I. A novel link between ran signal transduction and nuclear envelope proteins in plants. Plant Physiol. 2000;124:1507–10.
7. Li H, Wu G, Ware D, Davis KR, Yang Z. Arabidopsis Rho-related GTPases: differential gene expression in pollen and polar localization in fission yeast. Plant Physiol. 1998;118(2): 407–17.
8. Zhang Y, McCormick S. A distinct mechanism regulating a pollen-specific guanine nucleotide exchange factor for the small GTPase Rop in Arabidopsis thaliana. Proc Natl Acad Sci U S A. 2007;104(47):18830–5.
9. Duan Q, Kita D, Li C, Cheung AY, Wu HM. FERONIA receptor-like kinase regulates RHO GTPase signaling of root hair development. Proc Natl Acad Sci U S A. 2010;107(41): 17821–6.
10. Chang F, Gu Y, Ma H, Yang Z. AtPRK2 promotes ROP1 activation via RopGEFs in the control of polarized pollen tube growth. Mol Plant. 2013;6(4):1187–201.
11. Lemichez E, Wu Y, Sanchez JP, Mettouchi A, Mathur J, Chua NH. Inactivation of AtRac1 by abscisic acid is essential for stomatal closure. Genes Dev. 2001;15(14):1808–16.
12. Li H, Shen JJ, Zheng ZL, Lin Y, Yang Z. The Rop GTPase switch controls multiple developmental processes in Arabidopsis. Plant Physiol. 2001;126(2):670–84.
13. Baxter-Burrell A, Yang Z, Springer PS, Bailey-Serres J. RopGAP4-dependent Rop GTPase rheostat control of Arabidopsis oxygen deprivation tolerance. Science. 2002;296(5575): 2026–8.
14. Tao LZ, Cheung AY, Wu HM. Plant Rac-like GTPases are activated by auxin and mediate auxin-responsive gene expression. Plant Cell. 2002;14(11):2745–60.
15. Zheng ZL, Nafisi M, Tam A, Li H, Crowell DN, Chary SN, et al. Plasma membrane-associated ROP10 small GTPase is a specific negative regulator of abscisic acid responses in Arabidopsis. Plant Cell. 2002;14(11):2787–97.
16. Vernoud V, Horton AC, Yang Z, Nielsen E. Analysis of the small GTPase gene superfamily of Arabidopsis. Plant Physiol. 2003;131(3):1191–208.

17. Yamaoka S, Leaver CJ. EMB2473/MIRO1, an Arabidopsis Miro GTPase, is required for embryogenesis and influences mitochondrial morphology in pollen. Plant Cell. 2008;20(3): 589–601.
18. McElver J, Patton D, Rumbaugh M, Liu C, Yang LJ, Meinke D. The TITAN5 gene of Arabidopsis encodes a protein related to the ADP ribosylation factor family of GTP binding proteins. Plant Cell. 2000;12(8):1379–92.
19. Lu C, Zainal Z, Tucker GA, Lycett GW. Developmental abnormalities and reduced fruit softening in tomato plants expressing an antisense Rab11 GTPase gene. Plant Cell. 2001;13(8): 1819–33.
20. Yang Z. Small GTPases: versatile signaling switches in plants. Plant Cell. 2002;14(Suppl): S375–88.
21. Randazzo PA, Nie Z, Miura K, Hsu VW. Molecular aspects of the cellular activities of ADP-ribosylation factors. Sci STKE. 2000;2000(59):1.
22. Donaldson JG, Jackson CL. Regulators and effectors of the ARF GTPases. Curr Opin Cell Biol. 2000;12(4):475–82.
23. Geldner N, Friml J, Stierhof YD, Jurgens G, Palme K. Auxin transport inhibitors block PIN1 cycling and vesicle trafficking. Nature. 2001;413(6854):425–8.
24. Du C, Chong K. ARF-GTPase activating protein mediates auxin influx carrier AUX1 early endosome trafficking to regulate auxin dependent plant development. Plant Signal Behav. 2011;6(11):1644–6.
25. Pereira-Leal JB, Seabra MC. Evolution of the Rab family of small GTP-binding proteins. J Mol Biol. 2001;313(4):889–901.
26. Rutherford S, Moore I. The Arabidopsis Rab GTPase family: another enigma variation. Curr Opin Plant Biol. 2002;5(6):518–28.
27. Behnia R, Munro S. Organelle identity and the signposts for membrane traffic. Nature. 2005;438(7068):597–604.
28. Grosshans BL, Ortiz D, Novick P. Rabs and their effectors: achieving specificity in membrane traffic. Proc Natl Acad Sci U S A. 2006;103:11821–7.
29. Markgraf DF, Peplowska K, Ungermann C. Rab cascades and tethering factors in the endo-membrane system. FEBS Lett. 2007;581(11):2125–30.
30. Bos JL. Ras. In: Hall A, editor. GTPases. Oxford: Oxford University Press; 2000. p. 67–88.
31. Ridley A. Rho. In: Hall A, editor. GTPases. Oxford: Oxford University Press; 2000. p. 89–136.
32. Settleman J. Rac'n Rho: the music that shapes a developing embryo. Dev Cell. 2001;1:321–31.
33. Winge P, Brembu T, Kristensen R, Bones AM. Genetic structure and evolution of RAC-GTPases in Arabidopsis thaliana. Genetics. 2000;156(4):1959–71.
34. Zheng ZL, Yang Z. The Rop GTPase: an emerging signaling switch in plants. Plant Mol Biol. 2000;44(1):1–9.
35. Rossman KL, Der CJ, Sondek J. GEF means go: turning on RHO GTPases with guanine nucleotide-exchange factors. Nat Rev Mol Cell Biol. 2005;6(2):167–80.
36. Qiu JL, Jilk R, Marks MD, Szymanski DB. The Arabidopsis SPIKE1 gene is required for normal cell shape control and tissue development. Plant Cell. 2002;14(1):101–18.
37. Berken A, Thomas C, Wittinghofer A. A new family of RhoGEFs activates the Rop molecular switch in plants. Nature. 2005;436(7054):1176–80.
38. Gu Y, Li S, Lord EM, Yang Z. Members of a novel class of Arabidopsis Rho guanine nucleo-tide exchange factors control Rho GTPase-dependent polar growth. Plant Cell. 2006;18(2): 366–81.
39. Schlessinger J. Cell signaling by receptor tyrosine kinases. Cell. 2000;103(2):211–25.
40. Morris ER, Walker JC. Receptor-like protein kinases: the keys to response. Curr Opin Plant Biol. 2003;6(4):339–42.
41. Shiu SH, Bleecker AB. Receptor-like kinases from Arabidopsis form a monophyletic gene family related to animal receptor kinases. Proc Natl Acad Sci U S A. 2001;98(19):10763–8.
42. Kawasaki T, Koita H, Nakatsubo T, Hasegawa K, Wakabayashi K, Takahashi H, et al. Cinnamoyl-CoA reductase, a key enzyme in lignin biosynthesis, is an effector of small GTPase Rac in defense signaling in rice. Proc Natl Acad Sci U S A. 2006;103(1):230–5.

Chapter 3
Identification and Classification of Rho GTPases in Plants

Introduction

As found in animals, out of five classes of small GTPases, Ras family proteins have not been recognized in plants. However, Ras and Rho have only been known to be signaling-related proteins in yeast and animals while others play a role in cellular trafficking. In the signaling context, plant genomes are not endowed with many of G protein-coupled seven-transmembrane receptors, and virtually no receptor tyrosine kinases that are critical in animal signaling modules have been identified till date. Instead plants have evolved a dedicated class of receptor-like serine/threonine kinases (RLKs) for mediating the signal transduction. Recently, a single homolog of GPCR known as GCR1 was found in *Arabidopsis* involved in the regulation of ABA signaling in guard cells [1].

Another special class of signaling molecules of particular interest is Rho GTPases of plants (ROPs). They are specific to plants and have emerged as important signaling switch with respect to plants. The GTP hydrolysis-based switch has become the ideal control of signaling switch to control the external signal response [2]. The first Rho GTPase, Rho1Ps, was identified from pea plant in 1993; thereafter many such Rho GTPases have been identified in lower plants like mosses and higher plants. The *Arabidopsis* genome has been found to consist of 11 ROP genes encoding Rho-like proteins. These members of ROP were found to be similar to Rac-like proteins (70 % homology), rather than showing much similarity to Cdc42 and Rho of animals [3].

Nomenclature

Naming of ROP genes has been inconsistent in literature with several synonyms like AtRac, AtRho, AtRop in use. Subsequent to *Arabidopsis*, all the plant species have been found to contain multiple ROP proteins by different studies [4]. ROP

© The Author(s) 2015
G.K. Pandey et al., *GTPases*, SpringerBriefs in Plant Science,
DOI 10.1007/978-3-319-11611-2_3

Table 3.1 List of the ROP (Rho of plants) genes, their nomenclature, synonyms in literature, and the corresponding gene identification numbers

Gene name (Yang [4])	Synonyms	Locus ID
AtROP1	AtRAC11, Arac11	At3g51300
AtROP2	AtRAC4, Arac4	At1g20090
AtROP3	AtRAC1, Arac1	At2g17800
AtROP4	AtRAC5, Arac5	At1g75840
AtROP5	AtRAC6, Arac6, AtRac2	At4g35950
AtROP6	AtRAC3, Arac3, AtRac1	At4g35020
AtROP7	AtRAC2, Arac2	At5g45970
AtROP8	AtRAC9, Arac9	At2g44690
AtROP9	AtRAC7, Arac7	At4g28950
AtROP10	AtRAC8, Arac8	At3g48040
AtROP11	AtRAC10, Arac10	At5g62880

proteins of plants have drawn great attention because a single class of protein has evolved to perform the roles of a variety of Rho-related proteins like Cdc42, Rac, and Rho in controlling actin organization, cell polarity, and transduction of external signals. This also explains why ROP proteins have diversified themselves while some of the proteins showed overlapping functions, indicating plant cell requires numerous ROP proteins. Moreover, the mutants of ROP genes did not display obvious phenotype in forward-genetic screens indicating multiple genes involved in a single pathway and complementation of other ROPs in one's absence. Various ROP proteins and their synonyms have been listed in Table 3.1.

Small GTPase Complement in *Arabidopsis*

The full small GTPase complement in *Arabidopsis* was published in 2001 where the genome was scanned for the presence of G domain and the identified members were further classified into one of the small GTPase families [5]. This report grouped BLASTP-based found members into one of the accepted families of small GTPase proteins. According to this report, 93 genes were identified as small GTPase encoding genes in *Arabidopsis* classified as 11 members of Rho, 57 of Rab, 21 of Arf, and 4 members of Ran. While the majority of members belong to the Rab family of small GTPase, ROP (signaling GTPases) family members were equally diversified.

ROP family members were differentially expressed in many plant organs and growth stages. Given there are many members of ROP in plants, it is quite possible that they can serve as effective responsive switch, which is mediated by RAS and heterotrimeric G proteins in animals. The unique cycling features of small GTPases between two forms lead to the identification of their significant role in cell signaling. The class of mutants that was operative after these two active or inactive form is designated as DN dominant negative (inactive GDP-bound form) mutant and CA, constitutively active (active GTP-bound form) mutants. These mutant types were of tremendous significance for deciphering ROP GTPase function in plants. The greatest progress in understanding the role of ROP proteins in cell signaling came from the investigation of pollen tube growth. Pollen cell has long been considered as excellent single-cell-based study owing to its striking polar growth due to tip-based gradient of Ca^{2+} concentration. Supporting its role in polar growth is the localization of several Rop proteins in plasmalemma.

Furthermore, AtRop1 specific for pollen cells and AtRop6 specific for root hair cells were shown to induce depolarized growth. Most importantly, ROP related to animal and yeast Rho, Cdc42, Rac proteins is involved in cell polarity establishment.

ROP GTPase Complement in Rice

Rice small GTPases were first documented in a comparative analysis with *Arabidopsis*, human, Drosophila, and yeast [6]. As the map-based genome of rice was available since 2005, the continued effort of annotation is underway. The report on rice small GTPase was not based on locus-based analysis and hence many small GTPases have been misannotated by overrepresenting the numbers on some family. Hence there is a need to redefine this study on the rice small GTPase genes. RGAP (Rice Genome Annotation Project) was used as a focal point for all the gene identification processes. Keyword search was performed using "ras-related", "adp-ribosylation factor", "miro", "ras family domain containing protein" to fetch out small GTPase genes. With already 111 reported small GTPase genes in rice [6], based on the cDNA accession numbers, these data were also used to fetch out the corresponding locus IDs from the RGAP database using Basic Local Alignment Search Tool (BLAST). Genes that had only BAC clone positional information [6] were retrieved from the respective clones and BLAST searched in RGAP database to retrieve the corresponding locus IDs.

Furthermore, to enrich the collection, the HMM profile pattern was generated for more sequence similarity searches. Extensive database search for small GTPase genes in the rice genome yielded 85 genes (Fig. 3.1). The distribution of small GTPase genes among the four families is as follows: Rop, 9; Arf, 22; Rab, 43; Ran, 3; Miro, 4; unclassified, 4.

Among all, RAB constitutes the largest family, as in other eukaryotes followed by ARF, ROP, and RAN. In our study, we found that Jiang et al. [6] have reported five genes as multiple entries, which leads to the identification of higher numbers of genes in ARF and RHO families. This misinterpretation on the number of genes in the respective subfamily might be due to the lack of inclusion of information on actual BAC and cDNA clone by Jiang et al. [6]. This study also shed light on four novel small GTPase genes including one each of RAB, ARF, and ROP for the first time.

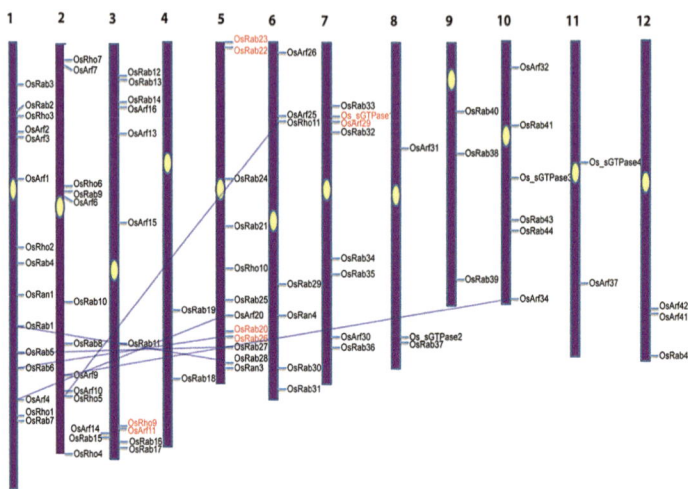

Fig. 3.1 Chromosomal distribution of small GTPases in rice. Jiang et al. [6] nomenclature was used to name the genes. However, novel genes identified were named as Os_sGTPase 1-4. Segmentally duplicated genes are connected by *blue lines* while the tandemly duplicated genes are marked in *red color* as consecutive *red characters*

GTP binding is a very common cellular functional activity and there are many proteins that fall into this criterion including initiation and elongation factors of protein synthesis besides α-subunit of heterotrimeric G protein. In the literature, small GTPase term is often also designated as small GTP-binding protein. Hence to account this, many GTP-binding proteins were initially retrieved and the number has almost gone up to 117 (data not shown). However, the domain prediction tools such as SMART, InterPro, and Pfam did not predict any functional GTPase domain in the additional members other than the reported 85 genes in this search. However, the additional members had GTP-binding activity. Also, there are many GTP-binding proteins that are too large in size (>30 kDa) than the universally designated size of GTPase to be called as small GTPase.

There is also another reason for concluding these 85 genes as small GTPase superfamily, which includes the result of phylogenetic tree of retaining additional members comprising GTP-binding domain. These additional members formed a separate clade because of evolutionary distance and sequence unrelatedness while the highly related typical small GTPases formed perfect clades among one another indicating similar evolutionary origin and close homology among themselves. These are some of the prerequisite characteristic features of small GTPase super-family that are fulfilled by the 85 reported genes but not by the additional gene family members in this analysis.

References

1. Pandey S, Assmann SM. The Arabidopsis putative G protein-coupled receptor GCR1 interacts with the G protein alpha subunit GPA1 and regulates abscisic acid signaling. Plant Cell. 2004;16(6):1616–32.
2. Zheng ZL, Yang Z. The Rop GTPase: an emerging signaling switch in plants. Plant Mol Biol. 2000;44(1):1–9.
3. Moshkov IE, Novikova GV. Superfamily of plant monomeric GTP-binding proteins: Rab proteins are the regulators of vesicles trafficking and plant responses to stresses. Russ J Plant Physiol. 2011;55(1):119–29.
4. Yang Z. Small GTPases: versatile signaling switches in plants. Plant Cell. 2002;14(Suppl): S375–88.
5. Vernoud V, Horton AC, Yang Z, Nielsen E. Analysis of the small GTPase gene superfamily of Arabidopsis. Plant Physiol. 2003;131(3):1191–208.
6. Jiang SY, Ramachandran S. Comparative and evolutionary analysis of genes encoding small GTPases and their activating proteins in eukaryotic genomes. Physiol Genomics. 2006;24(3): 235–51.

Chapter 4
Sequence, Structure, and Domain Analysis of GTPases in Plants

Introduction

Protein sequences ultimately determine the domain folding and proper structure formation for the rightful function of the GTPases. The sequence similarity among small GTPases is strikingly regular among the members of the superfamily. The sequence similarity of the small GTPase domain and conservation pattern can be deduced in plants to study the significance of such conservation in relation to the function. In addition, domain identification from its coded protein sequence is critical for the grouping of new small GTPase protein into an existing family. To begin with, *Arabidopsis* small GTPase proteins were classified into existing families by aligning to the already classified small GTPase proteins of yeast and human by ClustalW [1]. The phylogenetic analysis (neighbor-joining method) of small GTPases with yeast and human members revealed that they do not cosegregate with any of the Ras family proteins indicating they are evolutionarily divergent to the existing plant small GTPases [2].

Domain Identification and Confirmation of Rice Small GTPases

To verify the presence of small GTPase domain in the enriched gene collections, the protein sequences were mined out from the RGAP for entire corresponding locus IDs. Simple Modular Architecture Research Tool (SMART), InterPro, Pfam databases were used for domain confirmation. Moreover, all these databases act as filters in order to find the functional small GTPase. The list of small GTPases determined at the end of the analysis is presented in Table 4.1 along with the corresponding locus IDs.

© The Author(s) 2015
G.K. Pandey et al., *GTPases*, SpringerBriefs in Plant Science,
DOI 10.1007/978-3-319-11611-2_4

Table 4.1 Eighty-five small GTPase genes identified in the study with their RGAP locus identification number and synonym as used by Jiang et al. [2]

Locus Id	Synonym
LOC_Os01g62570	OsRho1
LOC_Os01g35850	OsRho2
LOC_Os01g12900	OsRho3
LOC_Os02g58730	OsRho4
LOC_Os02g50860	OsRho5
LOC_Os02g20850	OsRho6
LOC_Os02g02840	OsRho7
LOC_Os03g59590	OsRho9
LOC_Os05g43820	OsRho10
LOC_Os06g12790	OsRho11
LOC_Os01g47730	OsRab1
LOC_Os01g12730	OsRab2
LOC_Os01g08450	OsRab3
LOC_Os01g37800	OsRab4
LOC_Os01g51700	OsRab5
LOC_Os01g54590	OsRab6
LOC_Os01g62950	OsRab7
LOC_Os02g43690	OsRab8
LOC_Os02g21710	OsRab9
LOC_Os02g37420	OsRab10
LOC_Os03g46390	OsRab11
LOC_Os03g05280	OsRab12
LOC_Os03g05740	OsRab13
LOC_Os03g09140	OsRab14
LOC_Os03g60530	OsRab15
LOC_Os03g60870	OsRab16
LOC_Os03g62600	OsRab17
LOC_Os04g49530	OsRab18
LOC_Os04g39440	OsRab19
LOC_Os05g44050	OsRab20
LOC_Os05g27530	OsRab21
LOC_Os05g01490	OsRab22
LOC_Os05g01480	OsRab23
LOC_Os05g20050	OsRab24
LOC_Os05g38630	OsRab25
LOC_Os05g44070	OsRab26
LOC_Os05g46000	OsRab27
LOC_Os05g48980	OsRab28
LOC_Os06g35814	OsRab29
LOC_Os06g47260	OsRab30
LOC_Os06g50060	OsRab31
LOC_Os07g13530	OsRab32
LOC_Os07g09680	OsRab33

(continued)

Table 4.1 (continued)

Locus Id	Synonym
LOC_Os07g31370	OsRab34
LOC_Os07g33850	OsRab35
LOC_Os07g44040	OsRab36
LOC_Os08g41340	OsRab37
LOC_Os09g15790	OsRab38
LOC_Os09g35860	OsRab39
LOC_Os09g10940	OsRab40
LOC_Os10g14150	OsRab41
LOC_Os10g30520	OsRab43
LOC_Os10g31830	OsRab44
LOC_Os12g43550	OsRab45
LOC_Os01g23620	OsArf1
LOC_Os01g15010	OsArf2
LOC_Os01g16030	OsArf3
LOC_Os01g59790	OsArf4
LOC_Os02g22140	OsArf6
LOC_Os02g03610	OsArf7
LOC_Os02g47110	OsArf9
LOC_Os02g49980	OsArf10
LOC_Os03g59600	OsArf11
LOC_Os03g13860	OsArf13
LOC_Os03g59740	OsArf14
LOC_Os03g27450	OsArf15
LOC_Os03g10370	OsArf16
LOC_Os05g41060	OsArf20
LOC_Os06g12090	OsArf25
LOC_Os06g02390	OsArf26
LOC_Os07g12200	OsArf29
LOC_Os07g42820	OsArf30
LOC_Os08g15040	OsArf31
LOC_Os10g04580	OsArf32
LOC_Os10g42940	OsArf34
LOC_Os11g37640	OsArf37
LOC_Os12g38130	OsArf41
LOC_Os12g37360	OsArf42
LOC_Os01g42530	OsRan1
LOC_Os05g49890	OsRan3
LOC_Os06g39875	OsRan4
LOC_Os07g12170	Os_sGTPase1
LOC_Os08g41250	Os_sGTPase2
LOC_Os10g23100	Os_sGTPase3
LOC_Os11g19800	Os_sGTPase4

Multiple Alignment and Phylogenetic Analysis

In order to capture the whole evolutionary history of this superfamily in rice, the phylogenetic analysis was performed. The protein sequences of all the small GTPase genes identified were downloaded from the RGAP v6.1. These sequences were then aligned using stand-alone ClustalXv2.1 (For Mac OSX) software. The aligned output was used for further analysis. The phylogeny was constructed using the Molecular Evolutionary Genetics Analysis, MEGA v5.0, [3] software for Mac OSX. An un-rooted neighbor-joining tree was constructed using 1,000 bootstrap replications as test of phylogeny. The output was exported to Newick (New Hampshire tree) format for further viewing and editing in FigTree v1.3.1 software (Mac OSX version). The final output phylogenetic tree is shown in Fig. 4.1.

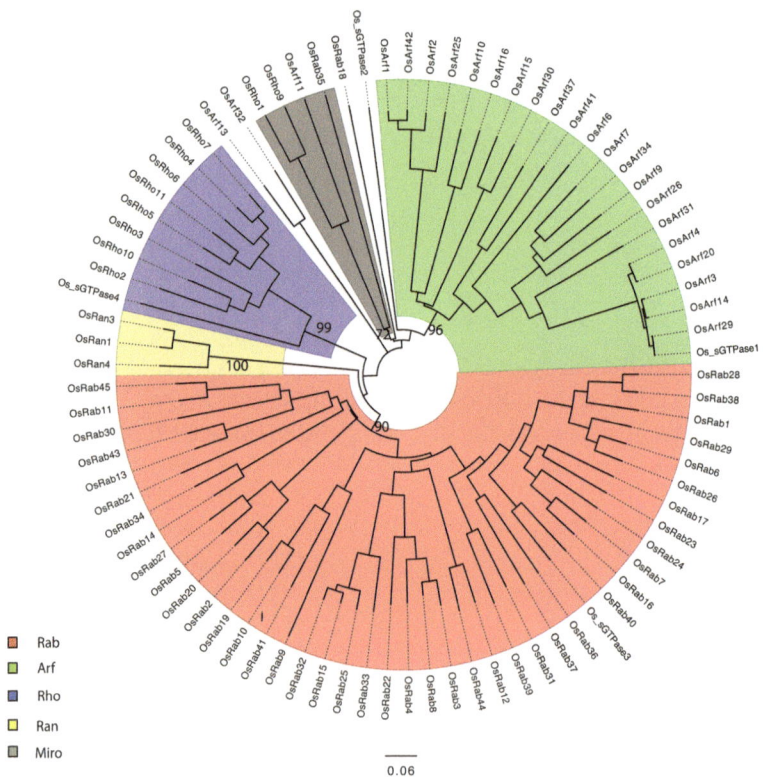

Fig. 4.1 Phylogenetic neighbor-joining tree of small GTPase superfamily, as deduced by aligning protein sequences in ClustalX and subsequently the tree construction in MEGA [3]. Bootstrapped for 1,000 times, the reliability for the test of phylogeny of the clades is marked with the maximum possible value of 100

Comparative Phyletic Analysis of Rice Small GTPase Genes

For the purpose of constructing the evolutionary history of rice small GTPase genes with other species like *Arabidopsis*, human, and yeast, the protein sequences of the other species excluding Hastings Research, Inc. were retrieved from the supplementary data of Yuksel et al. [4]. Each subfamily sequence of rice, *Arabidopsis*, human, and yeast was multiple aligned using ClustalX v2.1 and the aligned output was used for constructing an un-rooted neighbor-joining tree with MEGA v5.0 software [3]. All the subfamily comparative phylogenetic trees were bootstrapped for 1,000 replications for their reliability on the evolutionary history. The numbers of gene sequences used for the analysis are given in Table 4.2. The output of phylogeny of four distinct small GTPase families is shown in separate figures (Figs. 4.2, 4.3, 4.4, and 4.5).

Table 4.2 List of different number of GTPase subfamily members in four different species used for the phylogeny prediction

Organism	Families				
	Arf	Rab	Rop or Rho	Ran	Total
Oryza sativa	22	43	9	3	77
Arabidopsis thaliana	21	57	11	4	93
Homo sapiens	19	51	15	4	89
S. cerevisiae	6	11	5	2	24
Total	68	162	40	13	283

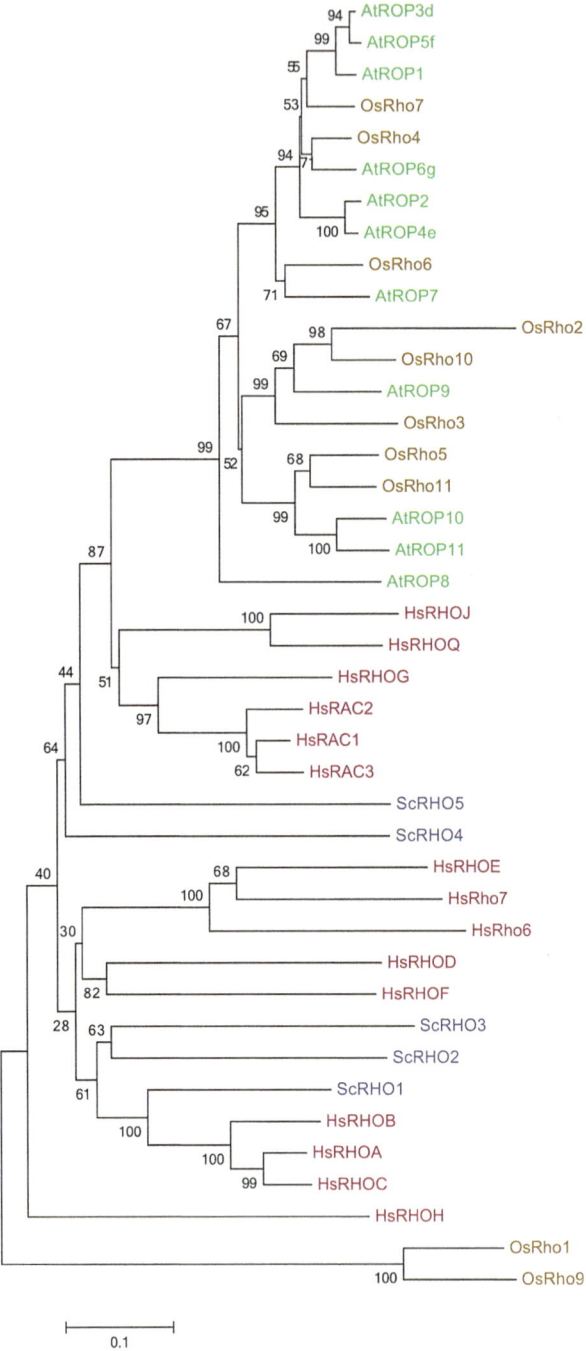

Fig 4.2 Comparative phylogenetic tree drawn from protein sequences of rice, *Arabidopsis*, human, and yeast Rho GTPases, color coded as *yellow, green, red,* and *blue*, respectively, by aligning with ClustalX program and subsequently an un-rooted neighbor-joining drawn with MEGA 5.0 program. Bootstrapped for 1,000 times, the maximum possible value is indicated in 100

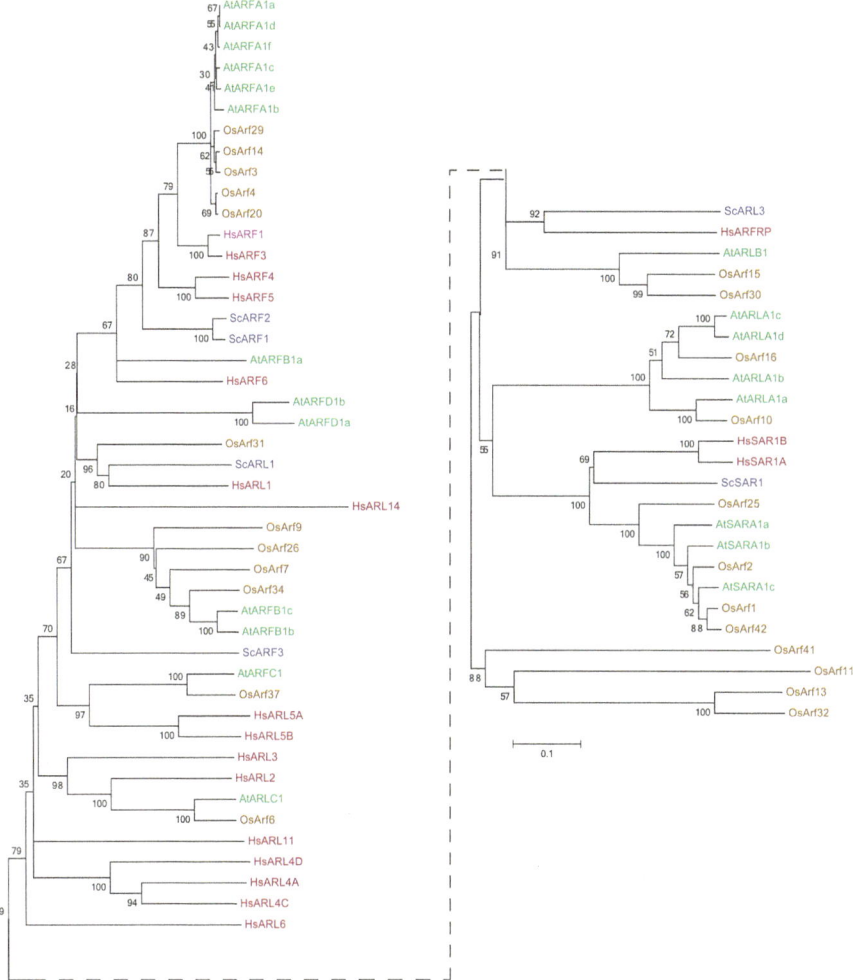

Fig. 4.3 Comparative phylogenetic tree of Arf GTPase constructed from protein sequences of four species namely rice, *Arabidopsis*, human, and yeast by aligning protein sequences in ClustalX followed by generating Neighbor-joining tree in MEGA 5.0. The color coding for four different species in the analysis of rice, *Arabidopsis*, human, and yeast are *yellow, green, red,* and *blue,* respectively. For convenience, the tree has been divided and connected by a *dotted line*

Fig. 4.4 Comparative
Rab-GTPase analysis among
four species namely rice,
Arabidopsis, human, and
yeast with color coding of
yellow, green, red, and *blue,*
respectively. The analysis
was performed by aligning
the protein sequences by
ClustalX and neighbor-
joining tree reported by
MEGA 5.0. For convenience,
the tree has been divided and
connected by a *dotted line*

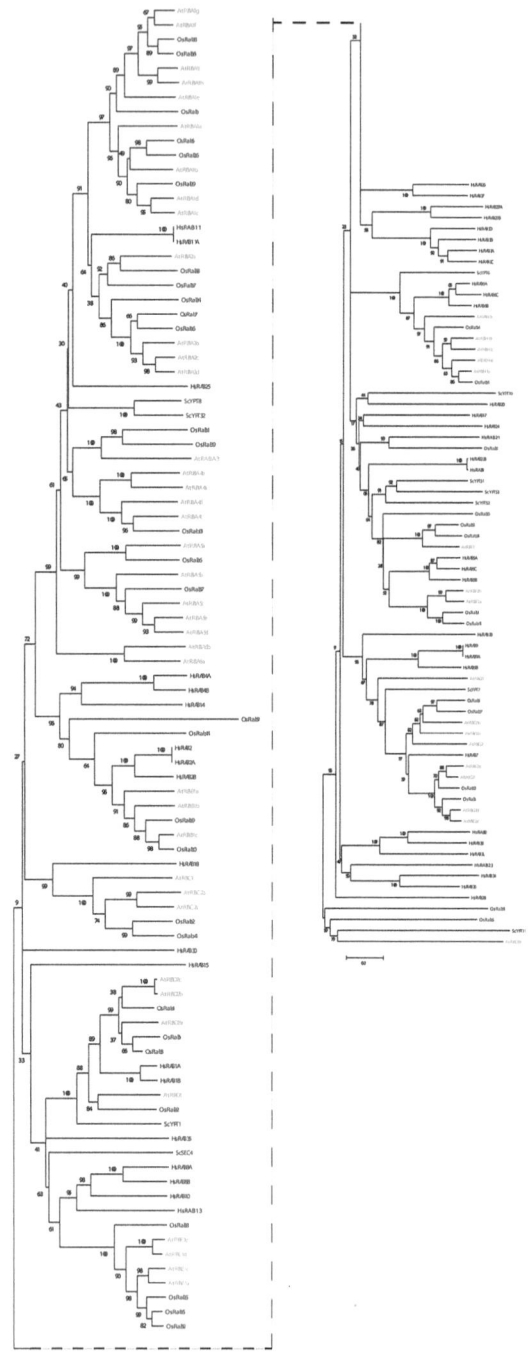

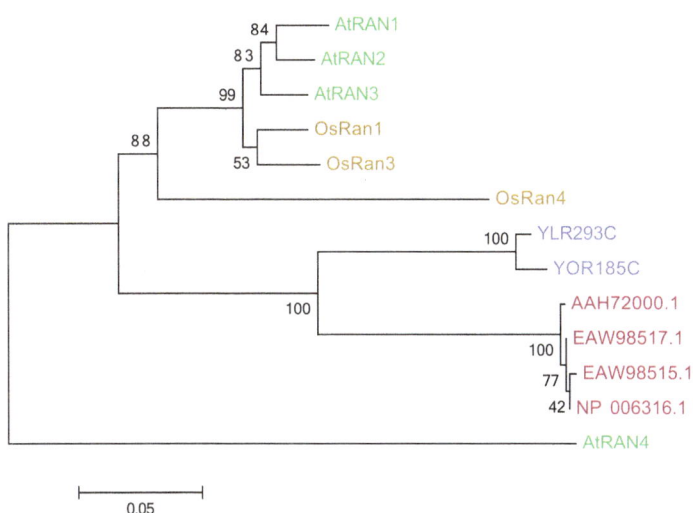

Fig. 4.5 Comparative phylogenetic tree from protein sequences of rice, *Arabidopsis*, human, and yeast Ran GTPases, color coded as *yellow, green, red,* and *blue*, respectively, by aligning with ClustalX program and subsequently un-rooted neighbor-joining tree drawn with MEGA 5.0 program. Bootstrapped for 1000 times, the maximum possible value is indicated in 100

Gene Nomenclature and Localization of Rice Small GTPase Complement

Small GTPase genes were named according to Jiang et al. [2]. Here, the four families were named as OsRho, OsRab, OsArf, and OsRan wherein Os stands for *Oryza sativa* and subsequently the names of typical four families of plant GTPases. The additional novel members based on the presence of functional domain were named as Os_sGTPase followed by a numerical indicating chromosomal order (1–4). All the small GTPase genes were positioned on their chromosomes and the segmental and tandem duplications were reported.

G Domain Conservation Pattern

WEBLOGO v2.8.2 application was used to generate the conserved residue logos by inputting a ClustalX aligned protein sequence for which the pattern is to be reported. This pattern was highlighted for all the subfamily level aligned sequences of rice, *Arabidopsis*, human, and yeast and the resulting conservation of G domains is presented in Fig. 4.6.

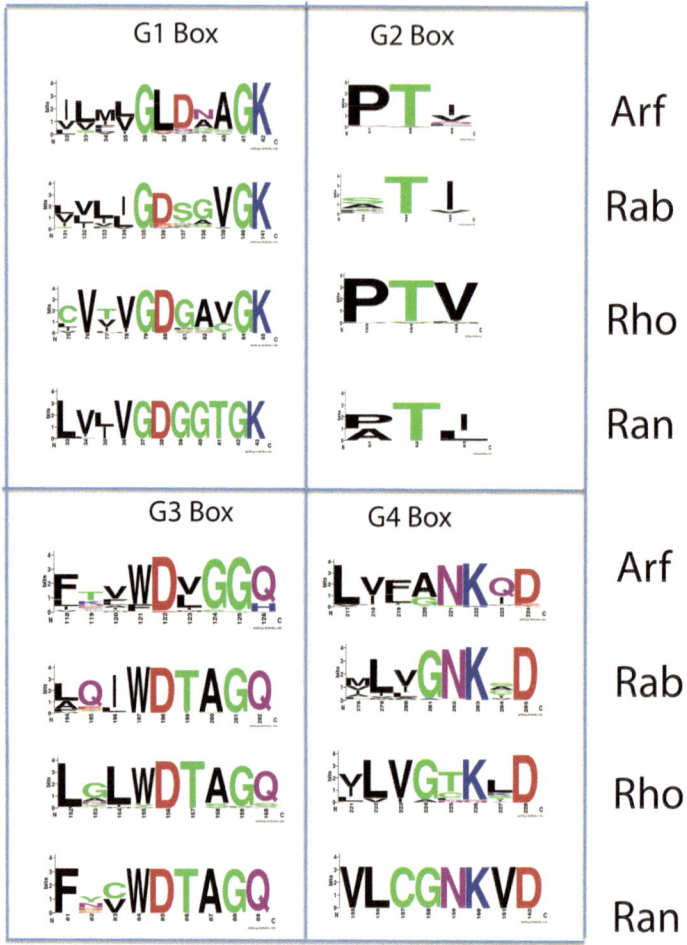

Fig. 4.6 The G1, G2, G3, and G4 domain conservation patterns of small GTPases as deduced by WEBLOGO using aligned protein sequences of Arf, Rab, Rho, and Ran families from four species namely rice, *Arabidopsis*, human, and yeast. The height of stack of amino acid residues indicates the degree of conservation on the given position

References

1. Vernoud V, Horton AC, Yang Z, Nielsen E (2003) Analysis of the small GTPase gene superfamily of Arabidopsis. Plant Physiol 131(3):1191–208
2. Jiang SY, Ramachandran S (2006) Comparative and evolutionary analysis of genes encoding small GTPases and their activating proteins in eukaryotic genomes. Physiol Genomics 24(3): 235–51
3. Tamura K, Peterson D, Peterson N, Stecher G, Nei M, Kumar S (2011) MEGA5: molecular evolutionary genetics analysis using maximum likelihood, evolutionary distance, and maximum parsimony methods. Mol Biol Evol 28:2731–2739
4. Yuksel B, Memon AR (2008) Comparative phylogenetic analysis of small GTP-binding genes of model legume plants and assessment of their roles in root nodules. J Exp Bot 59:3831–3844

Chapter 5
Expression of Small GTPases Under Stress and Developmental Conditions in Plants

Introduction

Regulation of gene expression is a classic response displayed by plants across different developmental stages. It shed light on a particular gene's role in correspondence with a stage of growth or a response to a hostile environment it is exposed at a given time. This process is both spatially and temporally regulated to fine-tune a gene's role during developmental stages apart from housekeeping genes. The plant-specific Rac/Rop small GTPases act as molecular switches in diverse signal transduction mechanisms [1, 2]. Several studies in different plant species have demonstrated the role of small GTPases for fine-tuning the stress and developmental responses [3–8]

It was of interest to determine whether the plant GTPases were expressed during stress, development, and phytohormone conditions, and, if they do, whether the different genes show distinct pattern. In order to understand their role in these conditions, we studied the expression pattern of the identified rice (85) and *Arabidopsis* (96) small GTPase gene families through Genvestigator (https://www.genevestigator. com) [9, 10], a Web-based search engine for gene expression. In particular, three abiotic stress conditions including drought, cold, and salinity were examined altogether for *Arabidopsis* and rice small GTPase gene families.

Expression Pattern of OsGTPases in Abiotic Stress

Differential expression has been a hallmark event for adaptive responses including wide variety of stress responses that plants encounter. In our analysis using public microarray data, we observed that under three different abiotic stress conditions only 20 genes were differentially expressed. Within these twenty, we found that many of the ROP genes were downregulated (*OsRho3*, *4*, *6*, and *11*) and most of

© The Author(s) 2015
G.K. Pandey et al., *GTPases*, SpringerBriefs in Plant Science,
DOI 10.1007/978-3-319-11611-2_5

Fig. 5.1 Heat map showing differentially expressed OsGTPases under *D*—drought, *C*—cold, *S*—salt stress conditions

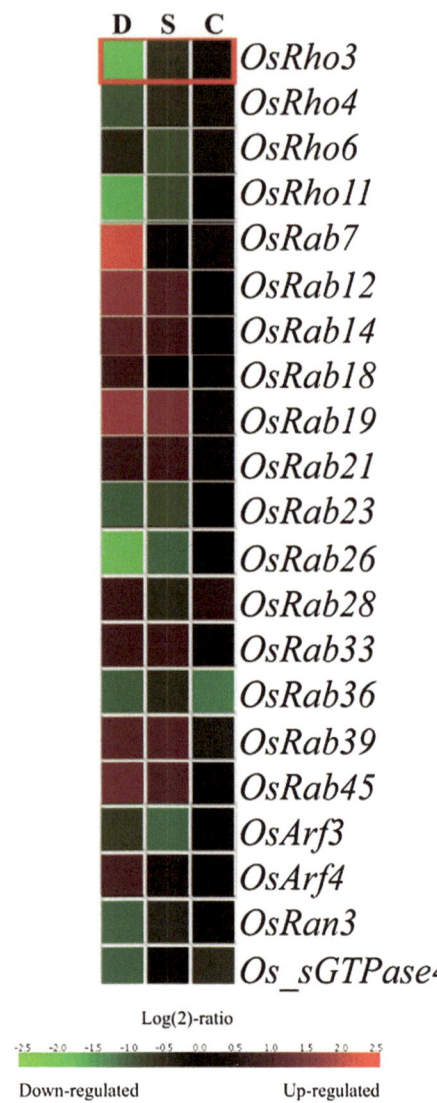

the Rab genes were either up- (*OsRab7, 12, 14, 18, 19, 21, 28, 33, 39,* and *45*) or down (*OsRab23, 26,* and *36*) regulated under salinity and water deficient conditions. ARF genes exhibited constant expression level, except for *OsArf3* exhibiting downregulation during salinity and drought conditions and *OsArf4* having elevated expression under drought stress. Single gene (*OsRan3*) was found to be downregulated during cold and drought stresses in the Ran family (Fig. 5.1).

Expression Pattern of OsGTPases in Developmental Stages

Microarray data from Genevestigator were also used to determine the expression patterns of rice GTPase in different developmental stages. The expression profile suggests a similar trend for ROP genes during seed and panicle developmental stages in four of the genes including *OsRho4*, *OsRho5*, *OsRho7*, and *OsRho11*. *OsRho6* shows an exceptionally high expression during seed germination stage with the extent of downregulation observed was manifold in later stages. *OsRho10* shows an overall downregulation during plant development. Genes in other families also exhibited a similar pattern of up- and downregulation in developmental stages. *OsRan1* and newly identified *Os_sGTPase3* were found to be induced under vegetative stages, whereas *Os_sGTPase2* expressed specifically under reproductive stages. *Os_sGTPase1* gets induced exclusively at maturation stage.

Among *OsArfs* seven genes, including, *OsArf3*, *OsArf6*, *OsArf10*, *OsArf16*, *OsArf26*, *OsArf30*, and *OsArf34* were found to be upregulated during seed germination stages. *OsArf14* showed exclusive induction during panicle developmental stage.

Amidst *OsRho* GTPases, *OsRho3*, *4*, and *7* showed upregulation in the vegetative stages, whereas single gene *OsRho6* was found to be solely induced during seed germination. The largest group of OsGTPases, *OsRABs,* shows varying expression pattern across all the developmental stages. Five genes (*OsRAB8*, *23*, *26*, *37*, and *45*) were found to be expressing differentially under both seed germination and panicle developmental stages. *OsRAB16*, *24*, *26*, *40*, and *43* were found to be altogether induced under initial stages of vegetative development, although, comparatively, fewer genes were found to be differentially expressed during reproductive developmental stages that includes *OsRAB1*, *17*, *19*, and *35* (Fig. 5.2).

Fig. 5.2 Heat map showing differential expression of OsGTPases in rice developmental stages

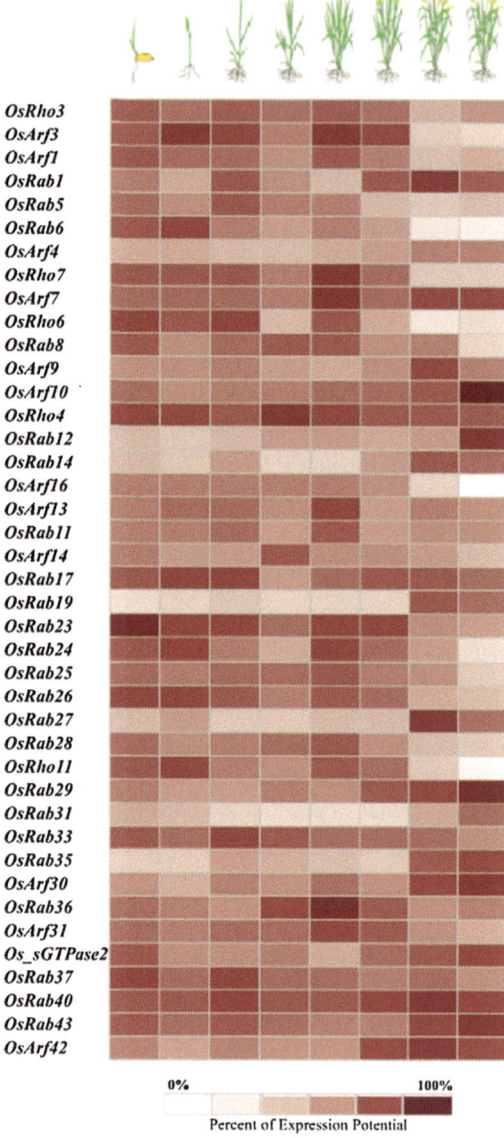

0%   100%

Percent of Expression Potential

Expression Pattern of OsGTPases During Phytohormone Treatment

Likewise, the expression profile of small GTPase complement in rice during hormonal conditions was extracted from Genevestigator. We found, among all, 23 genes were expressing differentially during ABA, and salicylic acid treatments.

Fig. 5.3 Heat map depicting differentially expressed OsGTPases under ABA (abscisic acid) and SA (salicylic acid)

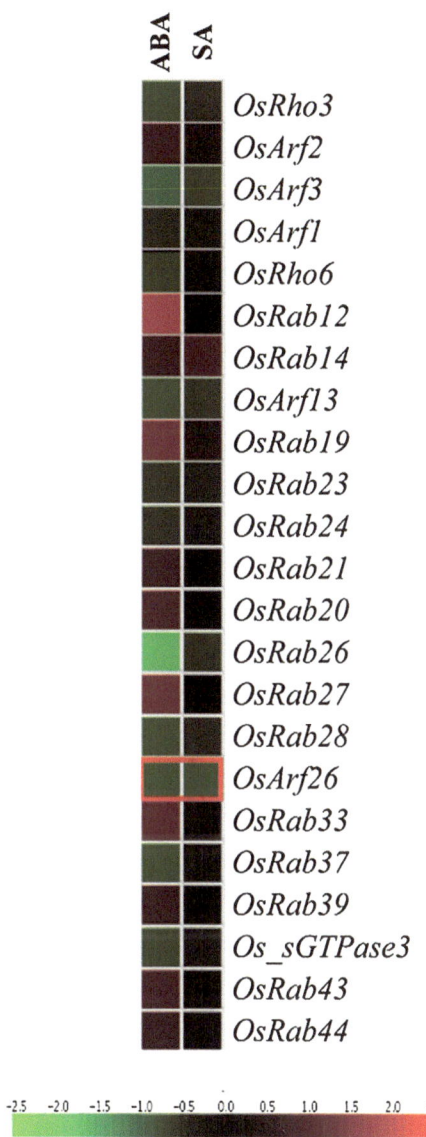

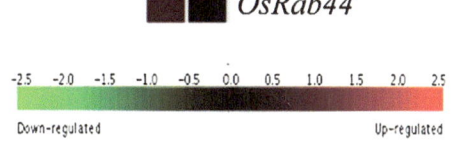

An elevated expression was observed for ten genes (*OsArf2*, *OsRab12*, *OsRab19*, *OsRab20*, *OsRab21*, *OsRab27*, *OsRab33*, *OsRab39*, *OsRab43*, and *OsRab44*) specifically during ABA and four (*OsRab1*, *OsRho6*, *OsRab37*, and *Os_sGTPase3*) were found to be downregulated exclusively in the same condition. Simultaneous upregulation in both ABA and SA was observed only for *OsRab14*, whereas nine genes (*OsRho3*, *OsArf3*, *OsRab1*, *OsArf13*, *OsRab23*, *OsRab24*, *OsRab26*, *OsRab28*, and *OsArf26*) were found to be concurrently downregulated (Figs. 5.3, 5.4, and 5.5).

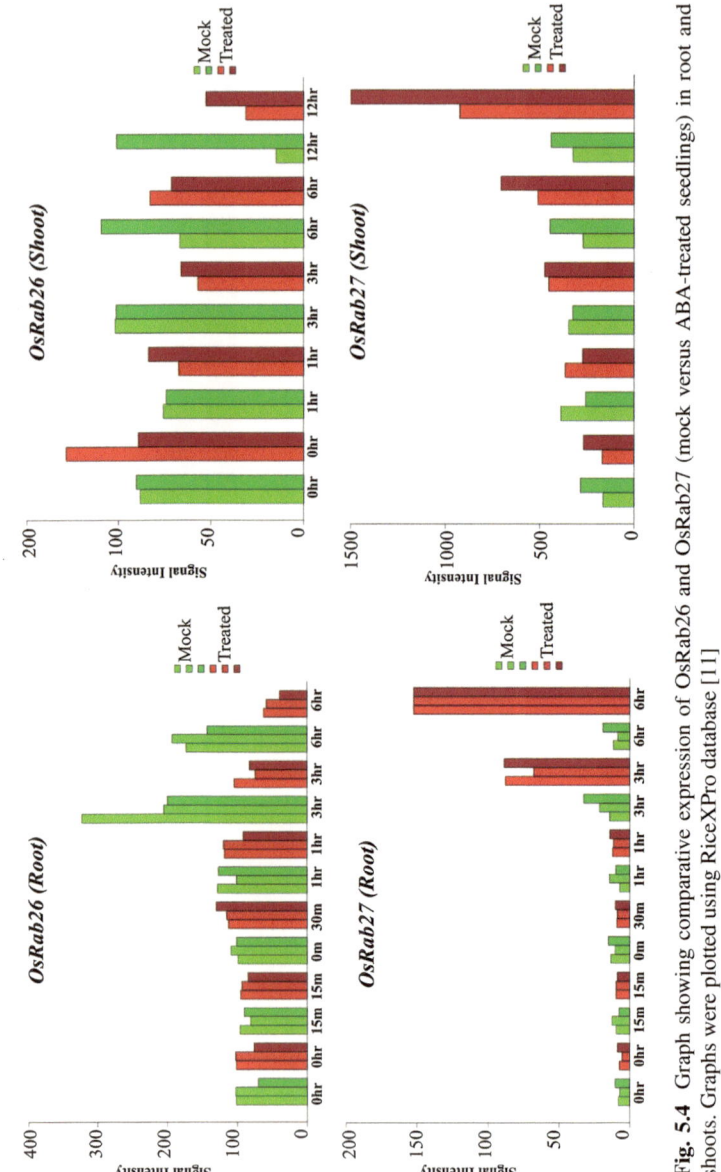

Fig. 5.4 Graph showing comparative expression of OsRab26 and OsRab27 (mock versus ABA-treated seedlings) in root and shoots. Graphs were plotted using RiceXPro database [11]

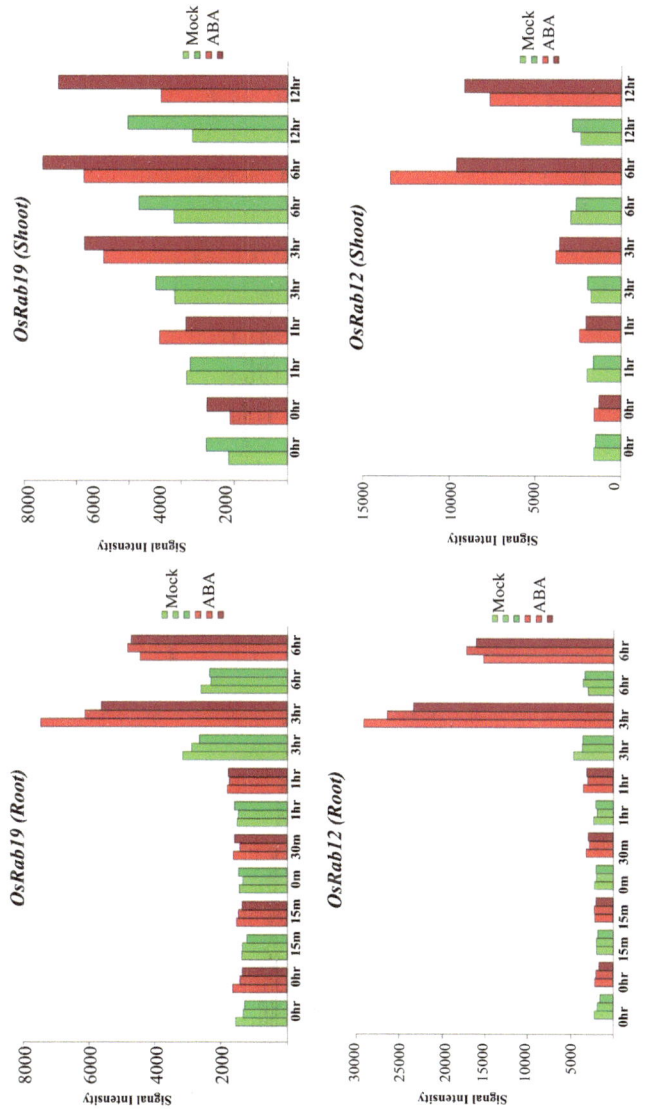

Fig. 5.5 Graph showing comparative expression of OsRab19 and OsRab12 (mock versus ABA treated seedlings) in root and shoots. Graph were plotted using RiceXPro database [11]

Expression Analysis of AtGTPases in Developmental Stages

Similarly, microarray data from Genevestigator were also used to determine the expression pattern of *Arabidopsis* GTPases gene family. The largest subgroup of AtGTPases, AtRABs, contained 57 members, and was found to exhibit a very broad expression pattern during plant development. *AtRABA2b* display very high expression during bolting stage followed by *AtRABA1i*, *AtRABA1h*, *AtRABA4d*, and *AtRABH1e*. Within *AtRABs*, highest expression was observed for *AtRABE1e* throughout the vegetative and reproductive developmental stages. In addition, both *AtRABH1c* and *AtRABA4c* did not express at all during developed rosette, bolting, and young flower stages of reproductive development. Among 11 AtROPs, eight were found to be upregulated commonly during young rosette stage. On the contrary, single gene *AtROP7* exhibited no expression in the same condition. Moreover, maximum expression under reproductive growth stages was observed for single gene *AtROP6g*. Further analysis of *AtArfs* detected steady expression pattern by all the genes across different developmental stages. Sole exception observed was *AtArfa1b*, having distinct higher expression during bolting stage of plant development. *AtARFD1b* and *AtARLA1b* were found not expressing during initial reproductive stages. Out of the four *AtRAN GTPases*, probeset was not available for one gene. The three inspected AtRAN genes were found to be expressed steadily in all the tested conditions of plant development (Fig. 5.6).

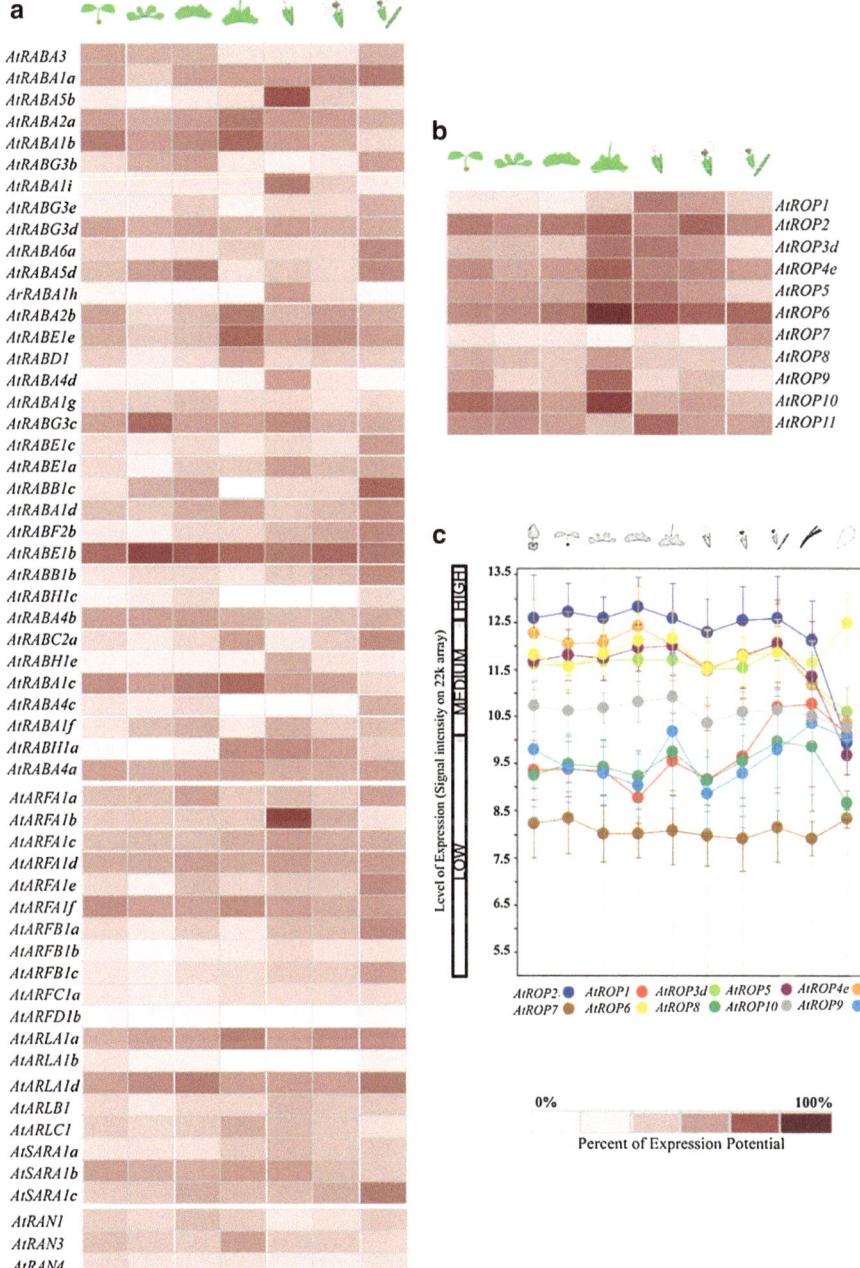

Fig 5.6 Heat map showing differential expression (**a**, **b**) of AtGTPases in different *Arabidopsis* developmental stages. (**c**) Expression of selected genes across different stages of plant development. For each stage, the expression values and standard deviations are calculated from all microarrays annotated for that particular stage

Expression Analysis of AtGTPases in Abiotic Stress

According to the expression profile derived from Genevestigator, fewer genes of *Arabidopsis* GTPases family were detected expressing differentially under abiotic stress conditions. In the largest group of AtRABs containing 57 members, 17 were found to be either up- or downregulated under cold, drought, and salt stress conditions.

We observed specific induction of six AtRABs (*AtRABA1a*, *AtRABA2d*, *AtRABB1b*, *AtRABC1*, *AtRABD2b*, and *AtRABH1c*) under osmotic stress, whereas nine genes (*AtRABA1b*, *AtRABA1c*, *AtRABA1g*, *AtRABA2a*, *AtRABA4b*, *AtRABA5b*, *AtRABA6a*, *AtRABG3d*, and *AtRABH1e*) were showing downregulation in the same stress condition. Three genes *AtRABA1c*, *AtRABC1*, and *AtRABG3d* were specifically showing upregulation during cold stress. During salt stress, nine of the AtRABs (*AtRABA1b*, *AtRABA2a*, *AtRABA2d*, *AtRABA4b*, *AtRABA5b*, *AtRABA6a*, *AtRABC1*, *AtRABG3d*, and *AtRABH1e*) were found to be downregulated, whereas upregulation was shown by four genes (*AtRABA2b*, *AtRABA4c*, *AtRABH1c*, and *AtRABA1g*) in the same condition. Even though several AtRABs were identified as salt and osmotic stress inducible, contrary to the reports no significant expression was observed for AtRABF1 during salt stress in both rice and *Arabidopsis* [4]. Relatedly, two of the AtROPs were showing downregulation when subjected to high salt conditions.

In the case of AtARFs, most of the genes were found to be responsive towards salt and osmotic stress conditions. Within 20 AtARFs, nine (*AtARFA1c*, *AtARFA1d*, *AtARFA1e*, *AtARFA1f*, *AtARFD1b*, *AtARLA1b*, *AtARLA1d*, *AtARLC1*, and *AtSARA1b*) exhibited downregulation under salinity. Remarkably, only *AtARFB1a* alone was upregulated in salinity and osmotic stress together. Among four of the AtRAN GTPases downregulation was caused by salinity in two of them, i.e., *AtRAN1d* and *AtRAN3d*. Although *AtRAN4* has been annotated as "salt inducible Ran1-like protein," interestingly, no modulation in expression was observed for this gene in any of the probed stress conditions (Fig. 5.7).

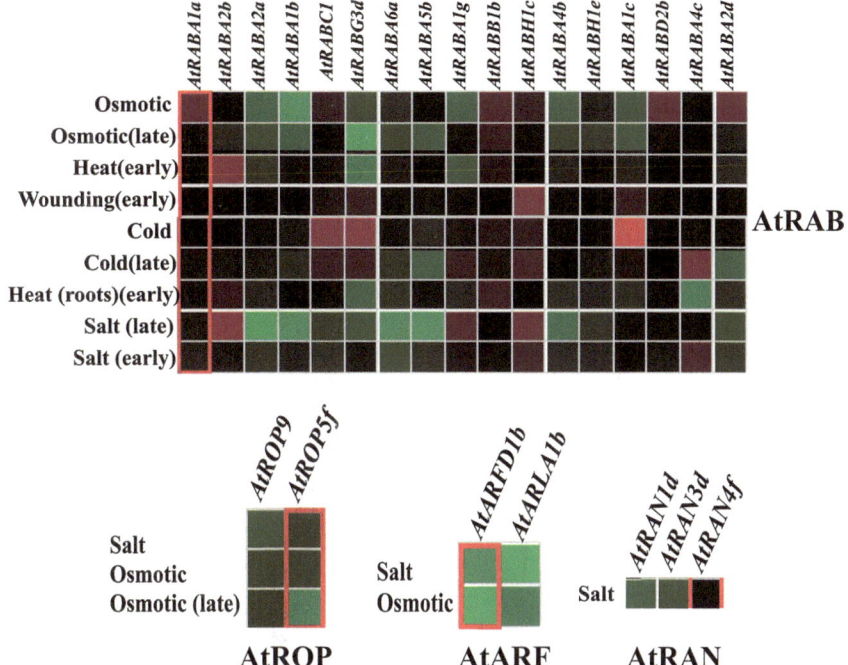

Fig. 5.7 Heat map of differentially expressed AtGTPases under abiotic stress conditions

Expression Analysis of AtGTPases During Phytohormone Treatment

The expression profile of *Arabidopsis* GTPases was also derived under different phytohormone treatments. Differential expression of GTPases was only observed during ABA treatment where 17 genes were found to be steadily expressed. Out of these 17, eight genes (ARA-2, ATSAR1, ATRabC2B, ATRab1C, ATRab2C, ATRabD2B, ATRabA2D, and ATRabA1F) showed elevated expression, while nine genes (ARA-1, ATARF1, ARA-4, ATRABA1G, ATRAB7D, ARLA1B, ATRABA1E, ATRABA1D, and ATRABA4A) displayed reduced expression (Fig. 5.8).

Fig. 5.8 Heat map showing differentially expressing AtGTPases under phytohormone ABA condition

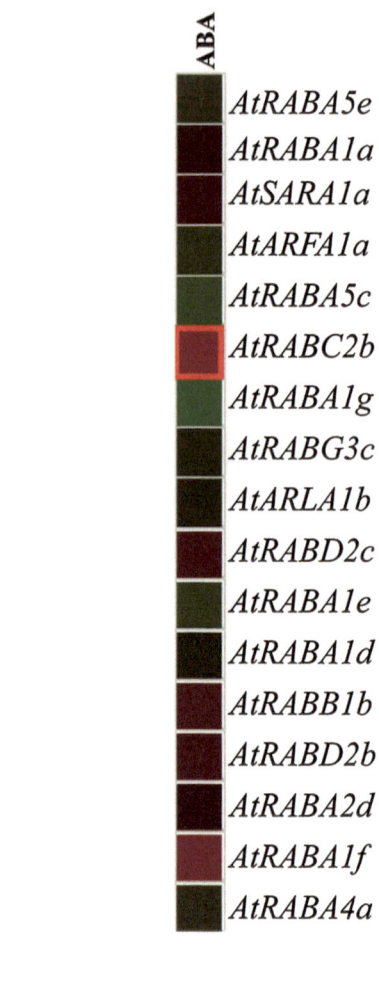

Log(2)-ratio

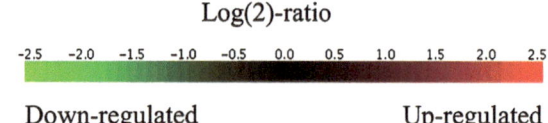

Down-regulated   Up-regulated

References

1. Nibau C, Wu HM, Cheung AY. RAC/ROP GTPases: 'hubs' for signal integration and diversification in plants. Trends Plant Sci. 2006;11(6):309–15.
2. Yang Z, Fu Y. ROP/RAC GTPase signaling. Curr Opin Plant Biol. 2007;10(5):490–4.
3. Kawasaki T, Henmi K, Ono E, Hatakeyama S, Iwano M, Satoh H, et al. The small GTP-binding protein rac is a regulator of cell death in plants. Proc Natl Acad Sci U S A. 1999;96(19):10922–6.

4. Bolte S, Schiene K, Dietz KJ. Characterization of a small GTP-binding protein of the rab 5 family in Mesembryanthemum crystallinum with increased level of expression during early salt stress. Plant Mol Biol. 2000;42(6):923–36.
5. Lemichez E, Wu Y, Sanchez JP, Mettouchi A, Mathur J, Chua NH. Inactivation of AtRac1 by abscisic acid is essential for stomatal closure. Genes Dev. 2001;15(14):1808–16.
6. Mazel A, Leshem Y, Tiwari BS, Levine A. Induction of salt and osmotic stress tolerance by overexpression of an intracellular vesicle trafficking protein AtRab7 (AtRabG3e). Plant Physiol. 2004;134(1):118–28.
7. Wong HL, Pinontoan R, Hayashi K, Tabata R, Yaeno T, Hasegawa K, et al. Regulation of rice NADPH oxidase by binding of Rac GTPase to its N-terminal extension. Plant Cell. 2007; 19(12):4022–34.
8. Poraty-Gavra L, Zimmermann P, Haigis S, Bednarek P, Hazak O, Stelmakh OR, et al. The Arabidopsis Rho of plants GTPase AtROP6 functions in developmental and pathogen response pathways. Plant Physiol. 2013;161(3):1172–88.
9. Vernoud V, Horton AC, Yang Z, Nielsen E. Analysis of the small GTPase gene superfamily of Arabidopsis. Plant Physiol. 2003;131(3):1191–208.
10. Zimmermann P, Hennig L, Gruissem W. Gene-expression analysis and network discovery using Genevestigator. Trends Plant Sci. 2005;10:407–9.
11. Sato Y, Takehisa H, Kamatsuki K, Minami H, Namiki N, Ikawa H, Ohyanagi H, Sugimoto K, Antonio B, Nagamura Y. RiceXPro Version 3.0: expanding the informatics resource for rice transcriptome. Nucleic Acids Res. 2013;41:D1206–13.

Chapter 6
Emerging Roles of Rho GTPases in Plants

Rho GTPases: Versatile Signaling Molecules in Plants

Small GTP (GTPase)-binding proteins are ubiquitous eukaryotic proteins acting as a binary switch cycling between GDP-bound inactive and GTP-bound active conformations. GTPases are small proteins ranging in molecular mass from 20 to 30 kDa performing diverse functions in plants and animals [1, 2]. Small GTPases are monomeric Guanine nucleotide-binding proteins closely related to α-subunit of heterotrimeric G proteins serving as master switch in the transmission of myriad of extracellular signals to intracellular pathways inside the cell. GTPases are generally localized at the plasma membrane permitting them to initiate signaling directly from the plasma membrane-associated receptors. Heterotrimeric GTPases perform important roles in plant signaling but their functions in plants are not as widespread as in animals [3, 4].

Small GTPases can be differentiated from the heterotrimeric G proteins on the basis of their regulatory mechanism. In response to an upstream factor, Guanine nucleotide exchange factors (GEFs) facilitate the activation of plasma membrane-associated GTPases by exchanging GDP-bound inactive form to GTP-bound active conformation. Subsequently, the activated GTPases interact with their downstream factors through its effector domain. The activated GTPases require GAPs (GTPase-activating protein) for their deactivation due to their weak intrinsic GTP hydrolysis property. Remarkably, plants do not have orthologs of animals' Rho family GTPases in their genome. They have been identified with a single large plant-specific subgroup of nearly identical small GTPases, termed Rop [5].

Plants have evolved novel ways to regulate and transmit ROP signals while conserving some of the common regulatory mechanisms present in animals. ROP GTPases control fundamental cellular mechanism in plants, such as cell polarity establishment in pollen tubes, root hair growth, cell morphogenesis, regulation of actin cytoskeleton, and hormonal responses, and have also been implicated in abiotic and biotic stress responses [6–9].

© The Author(s) 2015
G.K. Pandey et al., *GTPases*, SpringerBriefs in Plant Science,
DOI 10.1007/978-3-319-11611-2_6

Many studies have established that highly similar Rop proteins are also related at the functional level and are predicted to be acting redundantly. On the contrary, a single ROP can also regulate multiple processes by itself or in association with other functionally redundant ROPs. This mode of functioning by ROP genes tends to pose difficulty in their identification through forward genetics approach. There have been important advances in recent years that have revealed the fundamental mechanisms behind multifunctional regulation by relatively few ROP proteins in plants.

Rho Activates Plant Defense Mechanisms

Plant defense mechanisms are activated by the recognition of pathogen by disease resistance genes. These pathogen responsive genes activate a signaling pathway to develop resistance against that pathogen. Several studies have linked the function of ROPs as a molecular switch for plant defense responses and disease resistance.

In *Arabidopsis*, dual role for AtROP6 in developmental and pathogen response signaling was shown. The transcript levels of AtROP6 were found to be induced by auxin, and loss-of-function mutants had several developmental defects. AtROP6 was also found to be functionally associated with SA (salicylic acid)-mediated pathogen defense response [10].

Oryza sativa Rac/Rop GTPase, OsRAC1, has been shown as a positive regulator of ROI (reactive oxygen intermediates) production and hypersensitive response (HR) perhaps by interacting with the NADPH oxidase RbohB, ensuing resistance to pathogens [11, 12]. OsRAC1 forms a complex with RAR1 (required for Mla12 resistance) and HSP90 (Heat shock protein) to regulate innate immunity in rice [13]. Interestingly, subsequent studies in rice have failed to identify any additional positive regulator of blast resistance (Fig. 6.1).

On the contrary, OsRAC4 and 5 were identified as negative regulators of the same pathway while other Rac proteins were not found to be involved at all in disease resistance response [14]. In barley, the expression of activated Rho protein RACB enhanced susceptibility to *Blumeria graminis*, whereas microtubule-associated ROPGAP1 limits the plant susceptibility to penetration by this powdery mildew causing fungus [15]. Different reports in barley have identified three additional ROP proteins (HvRACB, HvRAC1, and HvRAC3) linked to both developmental and pathogen response [9, 15].

The exact mechanism by which ROP mediates barley resistance towards *B. graminis* is not yet fully understood. It is speculated that since *B. graminis* is a biotrophic pathogen required to penetrate into host cells for nutrition resulting in the invagination of host plasma membrane. This invasion of membrane is believed to be regulated by ROPs, possibly by secreting an invasion establishing factor [9].

The ortholog of barley *HvRACB*, in rice *OsRACB*, has been identified as a negative regulator of plant disease resistance pathway. The overexpression of plasma membrane-localized OsRACB renders plant susceptible to develop more chronic symptoms in response to blast pathogens [16]. The above evidences further

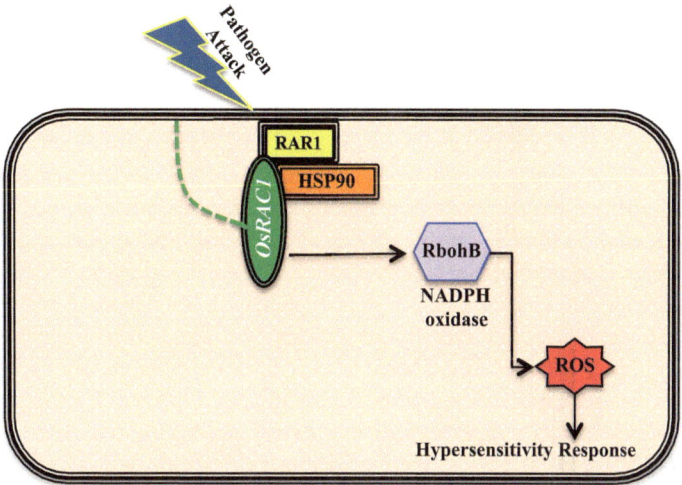

Fig. 6.1 OsRAC1 mediates innate immune response in rice. During pathogen attack (rice blast fungus) plasma membrane-associated OsRAC1, also known as immune switch, interacts with a cochaperone complex of RAR1 and HSP90 proteins to regulate downstream effector protein RbohB, an NADPH oxidase. NADPH oxidase-mediated ROS production and hypersensitivity response are required to combat pathogen attack in rice

strengthen the notion that ROPs act as versatile signaling molecules in plant defense responses.

In plants, a better understanding of the molecular roles of immune complexes containing various receptors and chaperones is needed to understand the role of Rho proteins in these signaling pathways.

Role in Intracellular Trafficking and Cell Polarity

Rho of plant (ROP) proteins function in multitude of regulatory pathways that include regulation of actin and cytoskeleton organization, vesicle trafficking, and cell polarity [17, 18]. Usually, vesicle trafficking begins at the plasma membrane by recruiting cargoes to form clathrin-coated vesicles that later undergo endocytosis [19]. Whereas intracellular vesicles progresse towards acceptor plasma membrane and subsequently undergo exocytosis to fuse with it [19].

The apical–basal distribution of PIN (PINFORMED) auxin transporters has been characterized in detail for their role in mediating auxin gradients during plant development [20]. PIN cargoes are conveyed by different endocytic/exocytic vesicles under the regulation of various small GTPases for their appropriate transport and polarization [21, 22]. Therefore, PIN regulation is regarded as the best molecular model to study ROP-mediated vesicle dynamics and plant-specific cell polarization.

Of the many small GTPase subfamily genes of ROPs in plants, a few of the genes control polarized growth and endomembrane trafficking events. With the help of coat protein complexes, they regulate vesicular transport between intracellular compartments by cytoskeleton arrangement and vesicle docking.

Plant ROP/RAC GTPases controlled F-actin structures and membrane trafficking regulates directional growth of cells [23, 24]. Plant cell expansion usually occurs in a diffuse manner, which is chiefly regulated by network of F-actin and filaments extending throughout the cytoplasm [25, 26]. Pollen tube and root hairs grow specifically at their apical end essentially in a polarized manner also known as tip growth. The fine filaments like F-actin underlying the plasma membrane were speculated to be involved in transport of secretory vesicles to their fusion sites along the plasma membrane [26, 27]. ROP/RAC GTPases certainly have important roles in the control of cell expansion and depolarization of diffuse cell growth [28, 29]. This has been confirmed in *Arabidopsis* where plants having mutation in SPIKE1 (SPK1) coding for ROP/RAC activating GEF protein had stunted growth and severe defects in polarized cell growth [30]. During tip growth in the cell, the typical distribution of ROP/RAC GTPases remains between cytoplasmic and plasma membrane specifically at the apex. The constitutive expression of ROP/RAC GTPases at the apical region depolarizes the cell growth and induces substantial swelling at this region. On the contrary, the loss of ROP/RAC GTPase activity inhibits tip growth [28, 31, 32].

In *Arabidopsis*, the function of AtROP11/RAC10 was analyzed in the root tip growth, where constitutive expression of this complex resulted in the depolarization of root hair growth, whereas wild-type expression caused swelling at the root hair tip without completely inhibiting apical tip growth of the cells [33]. Yet another link between ROP/RAC GTPase-mediated membrane trafficking was established by the direct interaction between *Arabidopsis* ICR1 (interactor of constitutively active ROP1) with Sec3, an exocyst component [34]. ICR1 is a coiled-coil scaffold protein specific to plants, whereas Sec3 is an established protein in the exocyst complex formation in yeast [35]. *Arabidopsis* genome has also been found encoding eight subunits of exocyst complex [36, 37]. Loss-of-function study of different Sec subunits in maize and *Arabidopsis* has demonstrated severe developmental defects such as inhibition of root hair growth and pollen tube elongation [37, 38].

In the case of Sec3, it was found that it does not interact directly with ROP/RAC GTPases; however, it readily associates with ICR1 to form a complex, which in turn is translocated to the plasma membrane [34]. Mutation in ICR1 compromises leaf morphogenesis as well as root development [34]. These observations establish that the regulation of exocyst function and membrane trafficking by ROP/RAC GTPases through their interaction with ICR1 is vital for polarized cell growth in plants [34].

Several downstream ROP/RAC effectors have also been reported to control cytoskeletal organization and polar cell growth. The precise balance between two antagonistic RIC1- and RIC4-dependent pathways mediate AtROP2- and AtROP4-directed pavement cell morphogenesis [29].

Role in Pollen Tube Growth

One of the key features of cells in multicellular organisms is that they are capable of migration and coordinate asymmetrical cell diffusion of cellular organelles, known as cell polarity [39]. Plants have developed an intricate network of polarized cell types to sustain axial growth, along which they grow and acclimatize to environmental conditions. Recent studies have provided evidence for the involvement of ROP GTPases and their interactors in the regulation of cell polarization. Auxin-mediated localization of ROP GTPases to specific membrane domains has been recognized in root hair cells [40]. Pollen tube growth occurs by the endocytosis and exocytosis of the vesicles at the extreme apical growing region [41].

Pollen tube elongation is governed by the gradual increase in ROP1 activity preferably towards the apical region of the pollen tube. Lateral propagation of the apical cap as a whole is prevented by RhoGDI and RhoGAP. In addition, ROP1 induces tip-localized actin microfilament formation that shoves ROP activators and inhibitors to the polar region, giving rise to both positive and negative feedback regulatory mechanisms [42]. Successively, a higher calcium gradient close to apical region is maintained by ROP1 for pollen tube elongation [43]. ROP1 regulates exocytosis by activating RIC3- and RIC4-mediated two different pathways. RIC4 stimulates organization of F-actin at the tip, while RIC3 stimulates the formation of cytosolic Ca^{2+} gradients needed for F-actin disassembly [44]. All the above components of ROP machinery are tightly connected to one another where the area of accumulation of each of them including ROP activation, actin accumulation, Ca^{2+} gradient formation coincides at the pollen tube apical region.

The genetic manipulation of ROP1 or its downstream regulators and effectors such as RIC3, RIC4, ICR1, and REN1 severely affects tip growth by causing depolarization [44–47]. Similar to other components, ROP-GAPs such as GAP1 and REN1 also accumulate at the apical region of pollen tube in order to restrict the lateral propagation of apical cells [48]. The distribution of REN1 in exocytic vesicles indicates the direct association of ROP signaling with vesicle docking, fusion, and transport. Thus, the overall control of ROP1 activity by feedback regulatory mechanism and exocytosis provides a competent manner for tip growth.

Role in Root Hair Development

Root hair plays an important role in plant development such as uptake of water and nutrient, plant anchorage, and association with microbes [49, 50]. Root hair grows in a polarized manner resulting from directional outgrowth of epidermal cells at a predefined region. Plant-specific Rop subfamily members play important roles in the regulation of root hair developmental processes. In *Drosophila*, Rac1 and Cdc42 are the two Rho GTPases involved in the regulation of wing hair formation, which corresponds to root hair development in plants [51]. The role of Rops has been speculated to be in the severing and rearrangement of F-actin filaments necessary or directional tip growth after trichoblast bulging [52]. The role of Rop in actin dynamics is more evident for the reason that Rop GTPase localized at the bud site reminiscent of the Cdc42 and Rho1p polar accumulation in yeast [53, 54]. Preliminary studies have shown that actin disrupting drugs did not impede polar localization of Rops whereas BFA did interfere with their localization at the future site of root hair formation. In *Arabidopsis*, BFA was found to disrupt the functioning of Arf GEF by causing endocytosis of the auxin efflux carrier PIN1 (Pin-formed-1) from its typical localization at the polar region on the plasma membrane [55]. This indicated a quintessential role of Arf-dependent vesicle trafficking in polarity induction in plant cells. Constitutive expression or overexpression of Rop GTPase at the root tip resulted in isotropic growth; in contrast, inactivation or removal of the same protein inhibits the polarized growth at the tip. In *Arabidopsis*, constitutive expression of AtRop4 and AtRop6 causes bulging in the hypocotyl cells in transformed plants suggesting a role of these protein in cell elongation [28].

A closely related member of Rop1, Rop2, was found to be expressing throughout the root hair development and shown to control polar site selection and root hair formation [32]. The overexpression of Rop2 in *Arabidopsis* resulted in the formation of profuse root hair with multiple tips while the overexpression of another Rop protein, Rop7, resulted in the inhibition of root hair growth. Thus, Rop2 functions as a positive regulator of root hair development, whereas Rop7 functions antagonistically. The same study also demonstrated that the overexpression of any other Rop from different subgroups had no apparent effect on root growth [32].

A distinct example of cross talk between environmental factors and ROP activity for the regulation of root tip growth was presented by Bloch et al. in 2011 [56, 57]. Previous studies on root growth have established the fact that a balance between fluctuation in pH and ROS distribution is necessary for tip growth. During root tip growth, the spatial regulation of Rop GTPases was assumed to coordinate these oscillations with respect to the environment [57]. The constitutive expression of *AtRop11* was shown to depolarize root hair growth [56]. Later on, more work on AtRop11 has established that the depolarization of root growth by its constitutive expression was also linked to the inhibition of ROS gradient inside the root apex. Additionally, bulging at the root tip is sensitive towards millimolar concentration of ammonium ions (NH_4^+) in the medium for the reason that it brings about pH fluctuation at the root tip [57].

Small GTPases Control Cell Morphogenesis

All multicellular organisms depend on cell and tissue morphogenesis for organ development. Growth of pavement cells in plant epidermal leaf cells is one of the well-studied example of morphogenic development [29]. Unlike animals, plants have distinct mechanism for planar cell polarity signaling pathway [40]. The arrangement of cortical microtubules and microfilaments in plants is controlled by

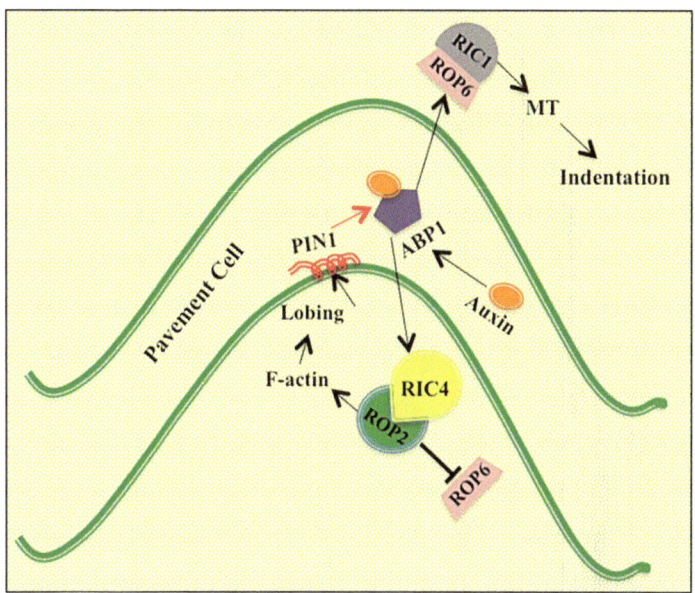

Fig. 6.2 Auxin regulates pavement cell interdigitation. ROP2 is localized towards lobing region to promote its protrusion, whereas ROP6 is localized towards indenting region for its growth. Auxin by activating ABP1 modulates pavement cell interdigitation. ROP2- and ROP6-mediated pathways are activated simultaneously to restrict PIN1 into the lobe apex which in turn activates a positive feedback loop consisting of auxin followed by ROP2, PIN1, and back to auxin

ROP GTPases [58, 59]. In *Arabidopsis*, the epidermal pavement cell morphogenesis is regulated by the countersignaling of two Rop-mediated antagonistic pathways. Rop2, on local activation, promotes localized outgrowth by activating RIC4-mediated cortical actin microfilament assembly. In the meantime, Rop2 also suppresses RIC1 mediated cortical microtubule organization. RIC1-mediated pathway also acts as a suppressor of activated ROP2 in the indentation zones. Thus, ROP2 regulated RIC1-MT inhibition and RIC4-MF promoting countersignaling pathways demarcate interdigitating separation of cortical domains between adjoining pavement cells [29]. The coordination between plant hormone auxin and specific Rop GTPases together organizes and restructures the cytoskeletal elements for cell morphogenesis and patterning [58] (Fig. 6.2).

Auxin binding protein1 (ABP1) has been identified as a auxin receptor that promptly activates cell expansion [60]. The auxin signaling acting downstream of ABP1 receptor promotes interdigitating pavement cell expansion by triggering Rop2 and Rop6 antagonistic pathways for the targeting of PIN1 proteins to the lobing regions of the plasma membrane [61].

RIC1 usually exists in association with microtubules and acts as effector molecule for ROP6. On activation by ROP6, RIC1 gets associated with microtubules and reorganizes them parallel to one another, which gives rise to indents in leaf PCs.

The ROP6-RIC1-mediated arrangement of cortical microtubules and cell morphogenesis represents the only well-studied pathway of cell morphogenesis in plants [29, 62]. Thus, the ROP-based signaling pathway that regulates lobe formation and pavement cell interdigitation is based on the initiation of a self-organizing signal. Auxin is considered to be one of the self-organizing signals. The interdigitating growth is severely inhibited in leaves with auxin biosynthesis deficiency [61].

Our current knowledge of RAC/ROP signaling is very limited, and perhaps a large multiprotein signaling cascade needs to be investigated. Till now, only a few RAC/ROPs and several of their interactor proteins have been characterized and a detailed research in this area might generate further insights into related signaling pathways in plants.

References

1. Lowy DR, Willumsen BM. Function and regulation of ras. Annu Rev Biochem. 1993;62: 851–91.
2. Terryn N, Van Montagu M, Inze D. GTP-binding proteins in plants. Plant Mol Biol. 1993;22(1):143–52.
3. Assmann SM. Heterotrimeric and unconventional GTP binding proteins in plant cell signaling. Plant Cell. 2002;14(Suppl):S355–73.
4. Ullah H, Chen JG, Temple B, Boyes DC, Alonso JM, Davis KR, et al. The beta-subunit of the Arabidopsis G protein negatively regulates auxin-induced cell division and affects multiple developmental processes. Plant Cell. 2003;15(2):393–409.
5. Zheng ZL, Yang Z. The RopGTPase: an emerging signaling switch in plants. Plant Mol Biol. 2000;44:1–9.
6. Zheng ZL, Nafisi M, Tam A, Li H, Crowell DN, Chary SN, et al. Plasma membrane-associated ROP10 small GTPase is a specific negative regulator of abscisic acid responses in Arabidopsis. Plant Cell. 2002;14(11):2787–97.
7. Tao LZ, Cheung AY, Wu HM. Plant Rac-like GTPases are activated by auxin and mediate auxin-responsive gene expression. Plant Cell. 2002;14(11):2745–60.
8. Baxter-Burrell A, Yang Z, Springer PS, Bailey-Serres J. RopGAP4-dependent Rop GTPase rheostat control of Arabidopsis oxygen deprivation tolerance. Science. 2002;296(5575):2026–8.
9. Schultheiss H, Dechert C, Kogel KH, Huckelhoven R. Functional analysis of barley RAC/ROP G-protein family members in susceptibility to the powdery mildew fungus. Plant J. 2003; 36(5):589–601.
10. Poraty-Gavra L, Zimmermann P, Haigis S, Bednarek P, Hazak O, Stelmakh OR, et al. The Arabidopsis Rho of plants GTPase AtROP6 functions in developmental and pathogen response pathways. Plant Physiol. 2013;161(3):1172–88.
11. Ono E, Wong HL, Kawasaki T, Hasegawa M, Kodama O, Shimamoto K. Essential role of the small GTPase Rac in disease resistance of rice. Proc Natl Acad Sci U S A. 2001;98(2): 759–64.
12. Wong HL, Pinontoan R, Hayashi K, Tabata R, Yaeno T, Hasegawa K, et al. Regulation of rice NADPH oxidase by binding of Rac GTPase to its N-terminal extension. Plant Cell. 2007;19(12):4022–34.
13. Thao NP, Chen L, Nakashima A, Hara S, Umemura K, Takahashi A, et al. RAR1 and HSP90 form a complex with Rac/Rop GTPase and function in innate-immune responses in rice. Plant Cell. 2007;19(12):4035–45.

14. Chen L, Shiotani K, Togashi T, Miki D, Aoyama M, Wong HL, et al. Analysis of the Rac/Rop small GTPase family in rice: expression, subcellular localization and role in disease resistance. Plant Cell Physiol. 2010;51(4):585–95.
15. Hoefle C, Huesmann C, Schultheiss H, Börnke F, Hensel G, Kumlehn J, et al. A barley ROP GTPase ACTIVATING PROTEIN associates with microtubules and regulates entry of the barley powdery mildew fungus into leaf epidermal cells. Plant Cell. 2011;23:2422–39.
16. Jung YH, Agrawal GK, Rakwal R, Kim JA, Lee MO, Choi PG, et al. Functional characterization of OsRacB GTPase-a potentially negative regulator of basal disease resistance in rice. Plant Physiol Biochem. 2006;44(1):68–77.
17. Nibau C, Wu HM, Cheung AY. RAC/ROP GTPases: 'hubs' for signal integration and diversification in plants. Trends Plant Sci. 2006;11(6):309–15.
18. Yang Z, Fu Y. ROP/RAC GTPase signaling. Curr Opin Plant Biol. 2007;10(5):490–4.
19. McMahon HT, Boucrot E. Molecular mechanism and physiological functions of clathrin-mediated endocytosis. Nat Rev Mol Cell Biol. 2011;12(8):517–33.
20. Grunewald W, Friml J. The march of the PINs: developmental plasticity by dynamic polar targeting in plant cells. EMBO J. 2010;29(16):2700–14.
21. Naramoto S, Kleine-Vehn J, Robert S, Fujimoto M, Dainobu T, Paciorek T, et al. ADP-ribosylation factor machinery mediates endocytosis in plant cells. Proc Natl Acad Sci U S A. 2010;107(50):21890–5.
22. Tanaka H, Kitakura S, Rakusova H, Uemura T, Feraru MI, De Rycke R, et al. Cell polarity and patterning by PIN trafficking through early endosomal compartments in Arabidopsis thaliana. PLoS Genet. 2013;9(5):e1003540.
23. Smith LG, Oppenheimer DG. Spatial control of cell expansion by the plant cytoskeleton. Annu Rev Cell Dev Biol. 2005;21:271–95.
24. Hussey PJ, Ketelaar T, Deeks MJ. Control of the actin cytoskeleton in plant cell growth. Annu Rev Plant Biol. 2006;57:109–25.
25. Dong CH, Xia GX, Hong Y, Ramachandran S, Kost B, Chua NH. ADF proteins are involved in the control of flowering and regulate F-actin organization, cell expansion, and organ growth in Arabidopsis. Plant Cell. 2001;13(6):1333–46.
26. Fu Y, Wu G, Yang Z. Rop GTPase-dependent dynamics of tip-localized F-actin controls tip growth in pollen tubes. J Cell Biol. 2001;152(5):1019–32.
27. Cardenas L, Lovy-Wheeler A, Kunkel JG, Hepler PK. Pollen tube growth oscillations and intracellular calcium levels are reversibly modulated by actin polymerization. Plant Physiol. 2008;146(4):1611–21.
28. Molendijk AJ, Bischoff F, Rajendrakumar CS, Friml J, Braun M, Gilroy S, et al. Arabidopsis thaliana Rop GTPases are localized to tips of root hairs and control polar growth. EMBO J. 2001;20(11):2779–88.
29. Fu Y, Gu Y, Zheng Z, Wasteneys G, Yang Z. Arabidopsis interdigitating cell growth requires two antagonistic pathways with opposing action on cell morphogenesis. Cell. 2005;120(5): 687–700.
30. Basu D, Le J, Zakharova T, Mallery EL, Szymanski DB. A SPIKE1 signaling complex controls actin-dependent cell morphogenesis through the heteromeric WAVE and ARP2/3 complexes. Proc Natl Acad Sci U S A. 2008;105(10):4044–9.
31. Kost B, Lemichez E, Spielhofer P, Hong Y, Tolias K, Carpenter C, et al. Rac homologues and compartmentalized phosphatidylinositol 4, 5-bisphosphate act in a common pathway to regulate polar pollen tube growth. J Cell Biol. 1999;145(2):317–30.
32. Jones MA, Shen JJ, Fu Y, Li H, Yang Z, Grierson CS. The Arabidopsis Rop2 GTPase is a positive regulator of both root hair initiation and tip growth. Plant Cell. 2002;14(4):763–76.
33. Bloch D, Monshausen G, Gilroy S, Yalovsky S. Co-regulation of root hair tip growth by ROP GTPases and nitrogen source modulated pH fluctuations. Plant Signal Behav. 2011;6(3): 426–9.
34. Lavy M, Bloch D, Hazak O, Gutman I, Poraty L, Sorek N, et al. A Novel ROP/RAC effector links cell polarity, root-meristem maintenance, and vesicle trafficking. Curr Biol. 2007; 17(11):947–52.

35. Zhang X, Orlando K, He B, Xi F, Zhang J, Zajac A, et al. Membrane association and functional regulation of Sec3 by phospholipids and Cdc42. J Cell Biol. 2008;180(1):145–58.
36. Elias M, Drdova E, Ziak D, Bavlnka B, Hala M, Cvrckova F, et al. The exocyst complex in plants. Cell Biol Int. 2003;27:199–201.
37. Hala M, Cole R, Synek L, Drdova E, Pecenkova T, Nordheim A, et al. An exocyst complex functions in plant cell growth in Arabidopsis and tobacco. Plant Cell. 2008;20(5):1330–45.
38. Wen TJ, Hochholdinger F, Sauer M, Bruce W, Schnable PS. The roothairless1 gene of maize encodes a homolog of sec3, which is involved in polar exocytosis. Plant Physiol. 2005;138(3):1637–43.
39. Orlando K, Guo W. Membrane organization and dynamics in cell polarity. Cold Spring Harb Perspect Biol. 2009;1(5):a001321.
40. Fischer U, Ikeda Y, Ljung K, Serralbo O, Singh M, Heidstra R, et al. Vectorial information for Arabidopsis planar polarity is mediated by combined AUX1, EIN2, and GNOM activity. Curr Biol. 2006;16:2143–9.
41. Lee YJ, Yang Z. Tip growth: signaling in the apical dome. Curr Opin Plant Biol. 2008;11(6): 662–71.
42. Yang Z, Lavagi I. Spatial control of plasma membrane domains: ROP GTPase-based symmetry breaking. Curr Opin Plant Biol. 2012;15(6):601–7.
43. Li H, Lin Y, Heath RM, Zhu MX, Yang Z. Control of pollen tube tip growth by a Rop GTPase-dependent pathway that leads to tip-localized calcium influx. Plant Cell. 1999;11(9): 1731–42.
44. Gu Y, Fu Y, Dowd P, Li S, Vernoud V, Gilroy S, et al. A Rho family GTPase controls actin dynamics and tip growth via two counteracting downstream pathways in pollen tubes. J Cell Biol. 2005;169(1):127–38.
45. Lee YJ, Szumlanski A, Nielsen E, Yang Z. Rho-GTPase-dependent filamentous actin dynamics coordinate vesicle targeting and exocytosis during tip growth. J Cell Biol. 2008;181(7): 1155–68.
46. Hwang JU, Vernoud V, Szumlanski A, Nielsen E, Yang Z. A tip-localized RhoGAP controls cell polarity by globally inhibiting Rho GTPase at the cell apex. Curr Biol. 2008;18(24): 1907–16.
47. Li S, Gu Y, Yan A, Lord E, Yang ZB. RIP1 (ROP Interactive Partner 1)/ICR1 marks pollen germination sites and may act in the ROP1 pathway in the control of polarized pollen growth. Mol Plant. 2008;1(6):1021–35.
48. Hwang JU, Wu G, Yan A, Lee YJ, Grierson CS, Yang Z. Pollen-tube tip growth requires a balance of lateral propagation and global inhibition of Rho-family GTPase activity. J Cell Sci. 2010;123(Pt 3):340–50.
49. Clarkson D. Factors affecting mineral nutrient acquisition by plants. Annu Rev Plant Physiol. 1985;36:77–115.
50. Bauer WD. Infection of legumes by rhizobia. Annu Rev Plant Physiol. 1981;32:407–49.
51. Eaton S, Wepf R, Simons K. Roles for Rac1 and Cdc42 in planar polarization and hair outgrowth in the wing of Drosophila. J Cell Biol. 1996;135(5):1277–89.
52. Baluska F, Salaj J, Mathur J, Braun M, Jasper F, Samaj J, et al. Root hair formation: F-actin-dependent tip growth is initiated by local assembly of profilin-supported F-actin meshworks accumulated within expansin-enriched bulges. Dev Biol. 2000;227(2):618–32.
53. Ziman M, Preuss D, Mulholland J, O'Brien JM, Botstein D, Johnson DI. Subcellular localization of Cdc42p, a Saccharomyces cerevisiae GTP-binding protein involved in the control of cell polarity. Mol Biol Cell. 1993;4(12):1307–16.
54. Yamochi W, Tanaka K, Nonaka H, Maeda A, Musha T, Takai Y. Growth site localization of Rho1 small GTP-binding protein and its involvement in bud formation in Saccharomyces cerevisiae. J Cell Biol. 1994;125(5):1077–93.
55. Steinmann T, Geldner N, Grebe M, Mangold S, Jackson CL, Paris S, et al. Coordinated polar localization of auxin efflux carrier PIN1 by GNOM ARF GEF. Science. 1999;286(5438): 316–8.

56. Bloch D, Lavy M, Efrat Y, Efroni I, Bracha-Drori K, Abu-Abied M, et al. Ectopic expression of an activated RAC in Arabidopsis disrupts membrane cycling. Mol Biol Cell. 2005;16(4):1913–27.

57. Bloch D, Monshausen G, Singer M, Gilroy S, Yalovsky S. Nitrogen source interacts with ROP signalling in root hair tip-growth. Plant Cell Environ. 2011;34(1):76–88.

58. Yang Z. Cell polarity signaling in Arabidopsis. Annu Rev Cell Dev Biol. 2008;24:551–75.

59. Fu Y, Li H, Yang Z. The ROP2 GTPase controls the formation of cortical fine F-actin and the early phase of directional cell expansion during Arabidopsis organogenesis. Plant Cell. 2002;14(4):777–94.

60. Badescu GO, Napier RM. Receptors for auxin: will it all end in TIRs? Trends Plant Sci. 2006;11(5):217–23.

61. Xu T, Wen M, Nagawa S, Fu Y, Chen JG, et al. Cell surface- and Rho GTPase-based auxin signaling controls cellular interdigitation in Arabidopsis. Cell. 2010;143:99–110.

62. Fu Y, Xu T, Zhu L, Wen M, Yang Z. A ROP GTPase signaling pathway controls cortical microtubule ordering and cell expansion in Arabidopsis. Curr Biol. 2009;19(21):1827–32.

Chapter 7
Cellular Localization of Small GTPases

Introduction

An alluring tendency of all organisms is their ability to compartmentalize subcellular signaling events. The coordinated assembly of signaling complexes at a specific subcellular location is generally a prerequisite for cellular responses such as cell division, migration, and polarization. In response to various chemical and growth ligands, small GTPases synchronize the cellular polarization and differentiation responses.

Intriguingly, the functional specificity in diverse organelles within the cell is achieved by lateral compartmentalization of the plasma membrane. The partitioning within plasma membrane coordinates cellular processes by spatially confining protein–protein and specific membrane lipid–protein interactions within the plasma membrane. Small GTPases readily associate with the plasma membrane through their hypervariable lipid domain. It has been proposed that both the hypervariable lipid domain and protein–protein interactions mediate subcellular targeting of GTPases.

The lateral segregation of macrodomains within the plasma membrane is critical for establishing cell polarity and largely determines the growth patterning of an organism. Plants are well adapted in modulating polarity establishment in cell growth, differentiation, and domain partitioning to reconstruct their body pattern throughout their life. GTP-binding proteins are the multifunctional signaling molecules, which when activated drive the lateral segregation of macrodomains at the plasma membrane. Activated RAC/ROP proteins act as molecular transducers in plants as they interact with downstream signaling molecules and are involved in cellular functions such as polarized tip growth in pollen cells by specifically localizing in cells such as pollen tubes and root tips [1–5].

Lately, the molecular mechanism controlling subcellular targeting of small GTPases has generated a great amount of interest among researchers. In this chapter, we will discuss the existing perceptions in subcellular targeting of small GTPases and how protein–protein interaction and lipid-derived plasma membrane association mediate their localization.

© The Author(s) 2015
G.K. Pandey et al., *GTPases*, SpringerBriefs in Plant Science,
DOI 10.1007/978-3-319-11611-2_7

Membrane Association of RAC/ROP GTPase

The plant small GTPase subfamilies (Rho/Rac, Rab, Arf, and Ran) are all key regulators in signaling pathways that control growth, differentiation, development, and defense responses. Upon activation, the GTPase induces and integrates intracellular signaling in eukaryotes. In addition to its activation, appropriate localization of the activated GTPase is also vital to its function particularly in processes involving cell division, migration, and polarity induction. The activation of ROP GTPase and its interaction with effector molecules generally entails its targeting to the membrane. Its consistent role in spatial regulation warrants regulated recruitment for proper function [1, 5–9]. Moreover, GTPase-mediated signaling is largely determined by its colocalization with the downstream regulators and effectors molecules within the same membrane enclosure.

Hence, targeting to specific membrane domains ensures their spatial control. During pollen tube elongation, the polarization of tip is coupled with the localization of activated G proteins to the apical region of the membrane [10]. Delocalization of RAC/ROPs to a different membrane domain severely compromises the polarized growth within pollen tubes. It is widely assumed that perhaps a regulating mechanism exists that ensures the suitable targeting of GTPases to the cell expansion sites where they are involved in the regulation of differential cell growth. The role of ROPs/RACs in regulating polar cell trafficking including polar auxin transporters has been recognized in *Arabidopsis*.

A ROP 1 interacting scaffold protein ICR1 (Interactor of Constitutive active ROP1) transports auxin transporters, PINs, to the polar domains at the plasma membrane. This implies that Rho-interacting protein ICR1 is crucial for directional auxin transport and dispersal for appropriate auxin-dependent pattern formation [11]. Therefore, the specific interaction of G proteins with regulators and effectors outlines their site of action leading to differential localization.

In *Nicotiana*, the interaction of *NtRAC5* with *NtRhoGDI1* has shown to be essential for the tip-restricted membrane localization of this protein. Any mutation in NtRAC5 inhibits its binding to NtRhoGDI1 resulting in its delocalization and consequently loss of depolarized pollen tube growth induction [12]. This suggests that plant GDIs are more or less responsible for the targeted localization of RAC/ROP GTPases akin to their animal counterparts [13].

Hypervariable Region Regulates RAC/ROP Localization

Different members of RAC/ROP GTPases are identical at the protein sequence level. The segregation of nearly identical GTPases can only be defined by their variable subcellular localization. This variability is largely determined by ~10 amino acid long hypervariable domain positioned prior to CAAX box.

In *Arabidopsis*, RAC/ROP GTPases are divided into two major subgroups, type-I and type-II, based on the type of lipid modification they accept on their hypervariable polybasic carboxyl terminal domain near the isoprenylation acceptor site [1, 14]. The functional difference between similar GTPases can be elucidated by their localization at different region in the cells, which is primarily determined by almost ten amino acids long hypervariable domain.

For lipid modification, type-I ROPs are assumed to be prenylated at their C-terminal by geranylgeranyltransferase, essentially for membrane attachment, while type-II ROPs undergo palmitoylation due to disruption of their C-terminal signature motif CAAL with an additional intron [15–17]. In flowering plants defect in protein prenylation does not appear to be lethal. However, a recent study showed the mechanism of protein prenylation and its role in mediating developmental processes in *P. patens*. The loss of Rab geranylgeranyltransferase activity causes severe phenotypes in *P. patens* than in *Arabidopsis* [18].

In plants, ROPs regulate the process of cell shape formation in one or different directions by either polar or diffused growth. Analogous to animals, prenylated Rho GTPase acts within cell membrane and cytosol as a result of its interaction with GDIs. Such a segregation between plasma membrane and cytosol perhaps allows a vital regulation of ROP recruitment to the site of action [6–9].

On the other hand, palmitoylation impedes interaction with GDIs, thereby dislodging them from cytosol activity. Primarily palmitoylation-dependent membrane localization was seen as the site of signaling for type-II RAC/ROP GTPases [19]. Thus, based on the different subcellular localization, the type-I and type-II ROPs coincide with their distinctive cellular function. A recent study in rice contributed another evidence to define the role of palmitoylation in membrane targeting. Small GTPase OsRAC1 regulates pathogen response against blast fungus by interacting with an NLR-type resistance (R) protein, Pit. Two palmitoylation sites in the N-terminal region of Pit specify the membrane localization of this protein. Mutation in these sites renders Pit to bind OsRac1 with lower affinity on the plasma membrane thereby failure of pit-mediated resistance to rice blast fungus. Thus, this study emphasizes the role of palmitoylation-dependent membrane localization and interaction in mediating disease resistance in rice [20].

Posttranslational Lipid Modifications Determine ROP Activity

The myriad of signaling responses mediated by ROP in plants necessitates specific regulation of their activity. Akin to animals, Guanine nucleotide exchange factor (GEFs), Guanine dissociation inhibitors (GDIs), and GTPase activating proteins (GAPs) together regulate the ROP activities in plants [12, 21–25].

However, it is unknown how ROPs concentrate themselves to the cytosol and plasma membrane and which cellular factors dictate their partition between the two compartments. This has been shown that the ROP localization to the root hair bud site is sensitive to protein trafficking inhibitor brefeldin A (BFA). In comparison to different actin disrupting drugs, BFA efficiently inhibits their polar localization to the trichoblast [1]. BFA constrains the enzyme activity of Arf GEF and thus strengthens the conviction that root hair initiation does not necessarily implicate actin, suggesting that vesicular trafficking or secreted proteins such as Arfs are crucial for the specialized transport of ROPs in roots [1]. ROPs require additive hydrophobic/lipid molecule for their activation and association with the membrane [4, 7]. Increasing evidence suggests that lipid microdomains or rafts in plasma membrane are likely to be the site for protein assembly and interactions in response to a particular stimulus. The nature of the lipid raft, also known as the detergent-resistant microdomains (DRMs), allows association only with acyl-modified proteins and not with prenylated proteins [26, 27].

This suggests that the type of posttranslational protein modification is sufficient to ascertain subcellular distribution, interaction, and function within the membrane. The *Arabidopsis* type-II RAC/ROP proteins associate with sterol-rich membrane domains, whereas palmitoylated type-I RAC/ROPs are assumed to be extricated from these domains [19, 28]. During cell signaling events type-I RAC/ROPs associate transiently with the sterol-rich membrane and can readily dissociate by depalmitoylation [19]. Thus, lipid modifications define the spatial segregation of RAC/ROP proteins into discrete domains to compartmentalize cellular processes and specify functions crucial in growth patterning within the plasma membrane.

In addition to the above-discussed fundamental regulation of ROPs activity via protein–protein interactions, subcellular localization, and membrane association, phosphorylation has been established to enhance the complexity of signaling pathway in plants. One of the phosphorylation motifs (S74) conserved in all plant ROPs is required for the nucleotide exchange and signaling activity of plant ROPs. Mutation in this potential phosphorylation site abolishes the binding of upstream PRONE domain-containing activator, ROPGEF protein and thus downstream GTPase activity. Based on the structural evidence, a notion has been presented that phosphorylation in plants could also determine ROP activation and signaling events [29].

Significance of Subcellular Localization in ROP Signaling

Although ROP family GTPases are remarkably homologous at the amino acid level, they show considerable functional specificity. Despite the conserved pathway of lipid modifications, ROP proteins demonstrate a great diversity of localizations. This variation in localization pattern perhaps contributes considerably to the functional diversity of these proteins. The variable localization pattern is largely determined by the type of lipid modification at the hypervariable region prior to CAAX motif.

The appropriate localization of RAC/ROP GTPases might be indispensable for many reasons.

The directed localization of these proteins to different subcellular compartments allows their specific interaction with various activators and effector molecules. Thus, function and localization of ROP proteins are correlated and can easily be anticipated from their membrane distribution.

References

1. Molendijk AJ, Bischoff F, Rajendrakumar CS, Friml J, Braun M, Gilroy S, et al. Arabidopsis thaliana Rop GTPases are localized to tips of root hairs and control polar growth. EMBO J. 2001;20(11):2779–88.
2. Molendijk AJ, Ruperti B. Palme K Small GTPases in vesicle trafficking. Curr Opin Plant Biol. 2004;7:694–700.
3. Vernoud V, Horton AC, Yang Z, Nielsen E. Analysis of the small GTPase gene superfamily of Arabidopsis. Plant Physiol. 2003;131(3):1191–208.
4. Lin YA, Wang Y, Zhu JK, Yang Z. Localization of a Rho GTPase implies a role in tip growth and movement of the generative cell in pollen tubes. Plant Cell. 1996;8:293–303.
5. Jones MA, Shen JJ, Fu Y, Li H, Yang Z, Grierson CS. The Arabidopsis Rop2 GTPase is a positive regulator of both root hair initiation and tip growth. Plant Cell. 2002;14(4):763–76.
6. Kost B, Lemichez E, Spielhofer P, Hong Y, Tolias K, Carpenter C, et al. Rac homologues and compartmentalized phosphatidylinositol 4, 5-bisphosphate act in a common pathway to regulate polar pollen tube growth. J Cell Biol. 1999;145(2):317–30.
7. Li H, Lin Y, Heath RM, Zhu MX, Yang Z. Control of pollen tube tip growth by a Rop GTPase-dependent pathway that leads to tip-localized calcium influx. Plant Cell. 1999;11(9):1731–42.
8. Fu Y, Li H, Yang Z. The ROP2 GTPase controls the formation of cortical fine F-actin and the early phase of directional cell expansion during Arabidopsis organogenesis. Plant Cell. 2002;14:777–94.
9. Lemichez E, Wu Y, Sanchez JP, Mettouchi A, Mathur J, Chua NH. Inactivation of AtRac1 by abscisic acid is essential for stomatal closure. Genes Dev. 2001;15(14):1808–16.
10. Yang Z. Small GTPases: versatile signaling switches in plants. Plant Cell. 2002;14(Suppl): S375–88.
11. Hazak O, Bloch D, Poraty L, Sternberg H, Zhang J, Friml J, et al. A rho scaffold integrates the secretory system with feedback mechanisms in regulation of auxin distribution. PLoS Biol. 2010;8(1):e1000282.
12. Klahre U, Becker C, Schmitt AC, Kost B. Nt-RhoGDI2 regulates Rac/Rop signaling and polar cell growth in tobacco pollen tubes. Plant J. 2006;46(6):1018–31.
13. Michaelson D, et al. Differential localization of Rho GTPases in live cells: regulation by hypervariable regions and RhoGDI binding. J Cell Biol. 2001;152:111–26.
14. Winge P, Brembu T, Kristensen R, Bones AM. Genetic structure and evolution of RAC-GTPases in Arabidopsis thaliana. Genetics. 2000;156(4):1959–71.
15. Ivanchenko M, Vejlupkova Z, Quatrano RS, Fowler JE. Maize ROP7 GTPase contains a unique, CaaX box-independent plasma membrane targeting signal. Plant J. 2000;24(1):79–90.
16. Lavy M, Bracha-Drori K, Sternberg H, Yalovsky S. A cell-specific, prenylation-independent mechanism regulates targeting of type II RACs. Plant Cell. 2002;14(10):2431–50.
17. Lavy M, Yalovsky S. Association of Arabidopsis type-II ROPs with the plasma membrane requires a conserved C-terminal sequence motif and a proximal polybasic domain. Plant J. 2006;46(6):934–47.
18. Thole JM, Perroud PF, Quatrano RS, Running MP. Prenylation is required for polar cell elongation, cell adhesion, and differentiation in Physcomitrella patens. Plant J. 2014;78(3):441–51.
19. Nibau C, Wu HM, Cheung AY. RAC/ROP GTPases: 'hubs' for signal integration and diversification in plants. Trends Plant Sci. 2006;11(6):309–15.
20. Kawano Y, Fujiwara T, Yao A, Housen Y, Hayashi K, Shimamoto K. Palmitoylation-dependent membrane localization of the rice R protein Pit is critical for the activation of the small GTPase OsRac1. J Biol Chem. 2014;289(27):19079–88. doi:10.1074/jbc.M114.569756.
21. Berken A, Thomas C, Wittinghofer A. A new family of RhoGEFs activates the Rop molecular switch in plants. Nature. 2005;436(7054):1176–80.
22. Gu Y, Li S, Lord EM, Yang Z. Members of a novel class of Arabidopsis Rho guanine nucleotide exchange factors control Rho GTPase-dependent polar growth. Plant Cell. 2006;18(2): 366–81.

23. Basu D, Le J, Zakharova T, Mallery EL, Szymanski DB. A SPIKE1 signaling complex controls actin-dependent cell morphogenesis through the heteromeric WAVE and ARP2/3 complexes. Proc Natl Acad Sci U S A. 2008;105(10):4044–9.
24. Wu G, Li H, Yang Z. Arabidopsis RopGAPs are a novel family of rho GTPase-activating proteins that require the Cdc42/Rac-interactive binding motif for Rop-specific GTPase stimulation. Plant Physiol. 2000;124(4):1625–36.
25. Hwang JU, Vernoud V, Szumlanski A, Nielsen E, Yang Z. A tip-localized RhoGAP controls cell polarity by globally inhibiting Rho GTPase at the cell apex. Curr Biol. 2008;18(24): 1907–16.
26. Mongrand S, Morel J, Laroche J, Claverol S, Carde JP, Hartmann MA, Bonneu M, Simon-Plas F, Lessire R, Bessoule JJ. Lipid rafts in higher plant cells: purification and characterization of Triton X-100-insoluble microdomains from tobacco plasma membrane. J Biol Chem. 2004; 279:36277–86.
27. Zacharias DA, Violin JD, Newton AC, Tsien RY. Partitioning of lipid-modified monomeric GFPs into membrane microdomains of live cells. Science. 2002;296:913–6.
28. Bloch D, Lavy M, Efrat Y, Efroni I, Bracha-Drori K, Abu-Abied M, et al. Ectopic expression of an activated RAC in Arabidopsis disrupts membrane cycling. Mol Biol Cell. 2005;16(4): 1913–27.
29. Fodor-Dunai C, Fricke I, Potocky M, Dorjgotov D, Domoki M, Jurca ME, et al. The phospho-mimetic mutation of an evolutionarily conserved serine residue affects the signaling properties of Rho of plants (ROPs). Plant J. 2011;66(4):669–79.

Chapter 8
Functional Genomic Perspective of Small GTPases

Introduction

A wide range of key cellular processes in eukaryotes that require the establishment of cellular homeostasis are governed by Rho GTPases. Contrary to animals and yeast, plants have been recognized with only one Rho GTPase subfamily known as Rho-like GTPases (ROPs). This suggests that plant must have evolved specific regulators and effectors in order to attain the extent of regulation needed for cellular developmental processes. Evidently, plants possess a combination of regulators to streamline different spatiotemporal cellular processes including morphogenesis, cell polarity, cell division, and endo- and exocytosis. Many studies have revealed that plants have evolved specific regulators, such as ROP-guanine exchange factors (GEFs) and the ROP interacting effectors to achieve the high level of regulation required for cellular processes. Some of the recent studies have shown that cross talk exists within distinct spatial and temporal functions of ROPs including actin dynamics, endo and exocytosis.

This chapter focuses on the proposed self-coordinating quality of ROPs in plants and how ROP-mediated cellular mechanisms are regulated and which effector and regulatory molecule contributes for this.

Regulatory Mechanism of Rho Signaling

Rho GTPases are small GTP-binding proteins of the Ras superfamily divided into five subfamilies in animals, i.e., Ras, Rab, Arf, Ran, and Rho groups. However, Ras subfamily proteins are altogether absent in plants putting them into a different class from animals. Since Ras proteins are principally involved in growth and developmental signaling pathways in animals entails that the similar pathways in plants might be regulated profoundly by different principles. 'Rho of plants' or ROP

© The Author(s) 2015
G.K. Pandey et al., *GTPases*, SpringerBriefs in Plant Science,
DOI 10.1007/978-3-319-11611-2_8

bearing signature GTP binding motif TKLD is found profusely in plants. An overwhelmingly high number and striking similarities among ROP proteins are the two key features conserved across monocots and dicots in plants. The large number of ROP proteins in plants suggests variable and vital functional roles of these proteins in plants. In spite of high sequence similarity, expression analysis shows ubiquitous expression for majority of them and specific localization was also predicted for a fair number of them. Genetic screens involving loss-of-function analysis suggest a small number of ROPs function redundantly.

The regulation of ROP activity presumably involves activation by coupling with GTP to stimulate downstream signaling and inactivation when bound to GDP after GTP hydrolysis. The activation of GTPases is mediated by Dbl-homology domain containing GEFs (GDP–GTP exchange factors) in animals. These exchange factors assist in the substitution of GDP for GTP on GTPases resulting in conformational changes and activation.

Intriguingly, none of the Dbl homology domains containing GEFs have been identified in plants. Earlier, a novel class of GEF has been identified in prokaryote (*Salmonella typhimurium*) that bears no resemblance to Dbl proteins [1] strengthening the notion that plants too might have evolved novel class of GEFs to activates ROPs. Preliminary study in *Lotus japonicus* has identified three putative GAP-like proteins that activates the ROPs, LjRac1 and LjRac2 [2]. Homology search in *Arabidopsis* also found LjGAP proteins encoded in its genome. RhoGEFs are multigenic characteristically membrane associated factors in plants essential for relaying signals from upstream regulators to downstream molecules in Rho signaling pathway [3, 4].

Apart from positive regulatory exchange factors, two different classes of proteins acting as negative regulators of Rho signaling have been identified [5]. By increasing their low intrinsic GTPase rate, RhoGAPs stimulate a GDP-bound confirmation change in GAPs (GTPase activating proteins) causing inactivation of Rho proteins [5]. Similarly, Rho suffers inactivation due to simultaneous inhibition of GDP–GTP exchange and targeted localization to cytoplasm from the plasma membrane by GDIs (Guanine nucleotide dissociation inhibitors) [6]. Due to this, the functional significance of RhoGAPs and RhoGDIs has long been considered inferior to GEFs. The posttranslational modification including S-acylation or prenylation of G-domain in RhoGTPase causing activation possibly prevents RhoGDI binding [7]. Remarkably, few RhoGTPases have been shown to interact with GDIs specifically when in an inactivated configuration; however, others show an autonomous interaction with RhoGDIs [8–11]. The evidence for this type of interaction was found in tobacco where preferential interaction of NtRhoGDI2 with inactive GDP-bound NtRac5 was demonstrated. NtRac5, RhoGTPase, regulates the pollen tube tip growth. Through yeast two-hybrid interaction analysis NtRac5 was found to be strongly interacting with the NtRhoGDI2 [10, 11]. RhoGDIs typically do not show any interaction with DN form of RhoGTPases, having weak affinity for both GTP and GDP. Subcellular localization analysis demonstrated cytoplasmic accumulation of NtRhoGDI2 in the pollen tube, whereas NtRac5 accumulates at the apical plasma membrane. The overexpression of either or both NtRhoGDI2 and NtRac5 severely compromises the pollen tube elongation. Intriguingly, the equivalent expression of both NtRhoGDI2 and NtRac5 neutralizes their effects and thus does not inhibit the pollen tube elongation. Another set of regulatory proteins RhoGDFs (RhoGDI dissociation factors) promotes dissociation of RhoGTPase/RhoGDI complexes and subsequent relocalization of RhoGTPases to the plasma membrane to facilitate their activation by RhoGEFs [12].

Studies in systems other than plants have shown that the regulation of Rho signaling by upstream regulators requires not only the activity control of RhoGEFs but also the direct regulation of interaction between RhoGTPase with RhoGAP and RhoGDI. Several essential cellular processes such as posttranslational modifications, protein binding, phosphorylation, and ubiquitinylation modify the RhoGAPs activity [5, 13, 14]. The interaction of RhoGTPase with RhoGDI also gets modulated by phosphorylation of any of these proteins by different protein kinases in response to signal-specific stimulus [12, 15, 16]. There are not many but few examples of phosphorylation-dependent regulation in plants including NtRhoGAP1. The relocation of NtRhoGAP1 to plasma membrane in pollen tube was suggested to be controlled by its phosphorylation-dependent interaction with an Nt14-3-3b-1 [9, 10].

Defying previous beliefs, various studies focusing on RhoGDI, RhoGAP activity regulation have defined their role in intricate fundamental cellular processes proposing them to be equally essential for the RhoGTPase activity as RhoGEFs [13]. Even though the RhoGTPases are functionally conserved proteins, plants have evolved specific downstream effectors to regulate intricate cellular processes. Apart from the conserved regulatory molecules including DHR2-type GEFs, typical RhoGAPs and RhoGDIs, plants have developed specific Guanine nucleotide exchange factors (RopGEFs).

Rho Interacting Proteins and Their Role in Plants

1. Regulators and Effectors of ROP
    A large number of varied RhoGTPase effector proteins are known in animals, such as kinases, actin and microtubule regulating formin proteins, and actin reorganizing WASP family of scaffolding proteins [17, 18]. Most of the effector protein from other systems are not found in plants, and in instances where homolog is present their role as effector protein has not been established. Interestingly, plants have evolved several families of plant-specific effector proteins for ROP signaling.
2. ROP Interactive Crib Domain Containing Proteins
    One of the first evidences for ROP effectors in plants includes a class of plant-specific CRIB (CDC42/Rac interactive binding) domain containing proteins also known as RICs (ROP Interactive CRIB motif containing proteins (RICs) [19]. The CRIB motif is responsible for the interaction of RICs with activated ROPs. Even though variations are observed in RICs at the sequence level, CRIB domain sequence is highly conserved in plants [20]. Similarly, a different group of plant ROP-interacting proteins having CRIB motif are known as the RopGAP proteins [19]. The CRIB motif sequences in plants were found to be highly homologous

to one another than in non-plant systems. This suggests that the CRIB motifs in plants perhaps originated from a single ancestor.

3. Two Counteractive Pathways Coordinate the Actin Dynamics in Pollen Tube Elongation

Genome-wide analysis in *Arabidopsis* has identified 11 highly divergent RICs to be encoded in its genome suggesting variable functional role for each of them in ROP signaling [19]. Subsequent functional characterization of various RICs in plants indicated that they certainly exhibit distinct functions. One of the best explored models of ROP-dependent actin dynamics is in the pollen tube tip growth. It has been established that three pollen-specific plant Rho GTPases (ROP1, 3, and 5) perform redundantly in the regulation of pollen tube tip growth [21]. Different studies in *Arabidopsis* and other plant species have revealed that the specific localization of ROPs in the apical plasma membrane region controls the dynamics of the tip F-actin and the creation of tip-directed calcium gradients. A well-defined counteractive mechanism involving single ROP is responsible for the tip actin dynamics. Two structurally distinct RICs (RIC3 and RIC4) coordinate with each other to control tip F-actin dynamics [22]. Further analysis of RIC functions revealed RIC4 as an effector of ROP2 and apparently possesses a conserved role in the actin assembly. Whereas the role of RIC3 was also turned out to be downstream of ROP2 in the accumulation of cytosolic Ca^{2+} at the apical region of pollen tube and subsequent disassembly of F-actin at the tip of the tubes [21]. Thus, both RIC3 and RIC4 act as effectors of ROP1 and coordinate counteractive pathways to regulate actin dynamics of pollen tube tip growth. In addition, RIC3- and RIC4-mediated antagonistic pathways also influence targeted exocytosis spatiotemporally to the apical plasma membrane for the extension of pollen tube [22].

4. The Antagonistic ROP2/RIC1 Pathway Promotes Microtubule Organization

Some of the recent advances in cell expansion studies have provided evidence that the morphogenesis of pavement cell is regulated by the countersignaling of two ROP-mediated pathways with contradictory effects on cell outgrowth.

ROP2-activated RIC4 stimulates the growth of interdigitating lobes and indentations by the assembly of cortical microtubules. At the same time ROP2 inactivates RIC1-mediated outgrowth-inhibiting antagonistic pathway to coordinate interdigitations amidst pavement cells [23, 24]. Interestingly, both ROP2 and ROP6 were found to be positively influenced by the plant hormone auxin. The coordinated activation of ROPs by auxin occurs instantly through a putative auxin receptor, auxin binding protein 1 (ABP1). These results suggest the role of RhoGTPase-based auxin-signaling mechanism in the coordinated control of cell expansion [25].

Unconventional Effectors of ROP/RAC GTPases

In plants, Rop/Rac GTPases perform a key regulatory role in diverse signaling pathways in cell growth, morphogenesis, and pathogen defense. The distinct function of GTP-bound ROP-containing complexes is specified by the associated binding partners, which directly influence the downstream effects of ROP activation. In *Arabidopsis*, one of the first reports has identified receptor-like cytoplasmic kinases and a cysteine-rich receptor kinases as molecular interactors of ROP complexes using yeast two-hybrid screening [26, 27]. In crop plant rice, different studies have identified several interactors of small GTP-binding protein OsRAC1. Earlier, OsRAC1 has been identified as a positive regulator of ROS (reactive oxygen species) production mediated by NADPH oxidase and cell death in plants to counter pathogen attack [28].

A different study has identified a key lignin biosynthesis enzyme OsCCR1 (*Oryza sativa* cinnamoyl-CoA reductase1) as an effector of OsRAC1. The OsCCR1 plays an important role in plant pathogen defense by producing an almost nondegrading mechanical barrier in the form of lignin polymer. Moreover OsCCR1 expression was found to be stimulated in response to a sphingolipid elicitor indicating its role in plant innate immunity [29]. The OsRAC1 immunity complex has multiple effector proteins. OsRACK1 was identified as yet another effector of this multiprotein immunity complex and is positively regulated by OsRAC1. Besides participating in the hormonal and developmental signaling pathways OsRACK1A also acts as a key regulator in the ROS production and plant immunity against rice blast infection [30]. Furthermore, a close functional link was also established between sphingolipid elicitor inducible mitogen-activated protein kinase 6 (MAPK6) and RAC1 in rice. Through coimmunoprecipitation assay, the close association between OsMAPK6 and active RAC1 has been proven [31]. Among all of the identified elicitors in the immune complex, the interaction of OsRAC1 with catalytic subunit protein Rboh is intriguing. Unlike animals, Rboh in plants contains an EF-hand motifs at its N-terminal, which is recognized by OsRAC1 for binding in a GTP-dependent manner. This interaction complex dissociates by the elevated Ca^{2+} flux in the cytosol mediated by calcium binding EF-hand motifs [32].

Plants have devised an altogether novel and specific effector proteins to regulate fundamental signaling pathways essential for cytoskeletal organization, vesicular trafficking, and other developmental processes. Study of these plant specific regulators and effector molecules have unravelled the novel prototype for the spatial control of small GTPases. Further elucidation of activation mechanism of plant specific effectors and regulators might provide molecular and cellular basis of the plant developmental and physiological processes regulated by GTPases.

References

1. Hardt WD, Chen LM, Schuebel KE, Bustelo XR, Galan JE. S. typhimurium encodes an activator of Rho GTPases that induces membrane ruffling and nuclear responses in host cells. Cell. 1998;93(5):815–26.
2. Borg S, Podenphant L, Jensen TJ, Poulsen C. Plant cell growth and differentiation may involve GAP regulation of Rac activity. FEBS Lett. 1999;453(3):341–5.
3. Berken A, Thomas C, Wittinghofer A. A new family of RhoGEFs activates the Rop molecular switch in plants. Nature. 2005;436(7054):1176–80.

4. Rossman KL, Der CJ, Sondek J. GEF means go: turning on RHO GTPases with guanine nucleotide-exchange factors. Nat Rev Mol Cell Biol. 2005;6(2):167–80.
5. Tcherkezian J, Lamarche-Vane N. Current knowledge of the large RhoGAP family of proteins. Biol Cell. 2007;99:67–86.
6. Hoffman GR, Nassar N, Cerione RA. Structure of the Rho family GTP-binding protein Cdc42 in complex with the multifunctional regulator RhoGDI. Cell. 2000;100(3):345–56.
7. Yalovsky S, Bloch D, Sorek N, Kost B. Regulation of membrane trafficking, cytoskeleton dynamics, and cell polarity by ROP/RAC GTPases. Plant Physiol. 2008;147:1527–43.
8. Ueda T, Kikuchi A, Ohga N, Yamamoto J, Takai Y. Purification and characterization from bovine brain cytosol of a novel regulatory protein inhibiting the dissociation of GDP from and the subsequent binding of GTP to rhoB p20, a ras p21-like GTP-binding protein. J Biol Chem. 1990;265:9373–80.
9. Klahre U, Becker C, Schmitt AC, Kost B. Nt-RhoGDI2 regulates Rac/ROP signaling and polar cell growth in tobacco pollen tubes. Plant J. 2006;46:1018–31.
10. Klahre U, Kost B. Tobacco RhoGTPase ACTIVATING PROTEIN1 spatially restricts signaling of RAC/ROP to the apex of pollen tubes. Plant Cell. 2006;18:3033–46.
11. Nomanbhoy TK, Cerione RA. Characterization of the interaction between RhoGDI and Cdc42Hs using fluorescence spectroscopy. J Biol Chem. 1996;271:10004–9.
12. DerMardirossian C, Bokoch GM. GDIs: central regulatory molecules in Rho GTPase activation. Trends Cell Biol. 2005;15:356–63.
13. Bernards A, Settleman J. GAP control: regulating the regulators of small GTPases. Trends Cell Biol. 2004;14:377–85.
14. Yoshida S, Pellman D. Plugging the GAP between cell polarity and cell cycle. EMBO Rep. 2008;9:39–41.
15. DerMardirossian C, Rocklin G, Seo JY, Bokoch GM. Phosphorylation of RhoGDI by Src regulates Rho GTPase binding and cytosol-membrane cycling. Mol Biol Cell. 2006;17:4760–8.
16. Qiao J, Holian O, Lee BS, Huang F, Zhang J, Lum H. Phosphorylation of GTP dissociation inhibitor by PKA negatively regulates RhoA. Am J Physiol Cell Physiol. 2008;295:C1161–8.
17. Bishop AL, Hall A. Rho GTPases and their effector proteins. Biochem J. 2000;348:241–55.
18. Perez P, Rincón SA. Rho GTPases: regulation of cell polarity and growth in yeast. Biochem J. 2010;426:243–53.
19. Wu G, Li H, Yang Z. Arabidopsis RopGAPs are a novel family of rho GTPase-activating proteins that require the Cdc42/Rac-interactive binding motif for rop-specific GTPase stimulation. Plant Physiol. 2000;124(4):1625–36.
20. Pirone DM, Carter DE, Burbelo PD. Evolutionary expansion of CRIB-containing Cdc42 effector proteins. Trends Genet. 2001;17(7):370–3.
21. Gu Y, Vernoud V, Fu Y, Yang Z. ROP GTPase regulation of pollen tube growth through the dynamics of tip-localized F-actin. J Exp Bot. 2003;54(380):93–101.
22. Gu Y, Fu Y, Dowd P, Li S, Vernoud V, Gilroy S, et al. A Rho family GTPase controls actin dynamics and tip growth via two counteracting downstream pathways in pollen tubes. J Cell Biol. 2005;169(1):127–38.
23. Fu Y, Gu Y, Zheng Z, Wasteneys G, Yang Z. Arabidopsis interdigitating cell growth requires two antagonistic pathways with opposing action on cell morphogenesis. Cell. 2005;120(5):687–700.
24. Fu Y, Xu T, Zhu L, Wen M, Yang Z. A ROP GTPase signaling pathway controls cortical microtubule ordering and cell expansion in Arabidopsis. Curr Biol. 2009;19(21):1827–32.
25. Xu T, Wen M, Nagawa S, Fu Y, Chen JG, Wu MJ, et al. Cell surface- and rho GTPase-based auxin signaling controls cellular interdigitation in Arabidopsis. Cell. 2010;143(1):99–110.
26. Molendijk AJ, Ruperti B, Singh MK, Dovzhenko A, Ditengou FA, Milia M, et al. A cysteine-rich receptor-like kinase NCRK and a pathogen-induced protein kinase RBK1 are Rop GTPase interactors. Plant J. 2008;53(6):909–23.
27. Dorjgotov D, Jurca ME, Fodor-Dunai C, Szucs A, Otvos K, Klement E, et al. Plant Rho-type (Rop) GTPase-dependent activation of receptor-like cytoplasmic kinases in vitro. FEBS Lett. 2009;583(7):1175–82.

28. Kawasaki T, Henmi K, Ono E, Hatakeyama S, Iwano M, Satoh H, et al. The small GTP-binding protein rac is a regulator of cell death in plants. Proc Natl Acad Sci U S A. 1999;96: 10922–6.
29. Kawasaki T, Koita H, Nakatsubo T, Hasegawa K, Wakabayashi K, Takahashi H, et al. Cinnamoyl-CoA reductase, a key enzyme in lignin biosynthesis, is an effector of small GTPase Rac in defense signaling in rice. Proc Natl Acad Sci U S A. 2006;103(1):230–5.
30. Nakashima A, Chen L, Thao NP, Fujiwara M, Wong HL, Kuwano M, et al. RACK1 functions in rice innate immunity by interacting with the Rac1 immune complex. Plant Cell. 2008; 20(8):2265–79.
31. Lieberherr D, Thao NP, Nakashima A, Umemura K, Kawasaki T, Shimamoto K. A sphingo-lipid elicitor-inducible mitogen-activated protein kinase is regulated by the small GTPase OsRac1 and heterotrimeric G-protein in rice 1[w]. Plant Physiol. 2005;138(3):1644–52.
32. Wong HL, Pinontoan R, Hayashi K, Tabata R, Yaeno T, Hasegawa K, et al. Regulation of rice NADPH oxidase by binding of Rac GTPase to its N-terminal extension. Plant Cell. 2007; 19(12):4022–34.

Chapter 9
Systemic Approaches to Resolve Spatiotemporal Regulation of GTPase Signaling

Introduction

ROP/RAC family GTPases are the pivotal signaling proteins conserved in multicellular organisms. In plants, small GTPases ROP/RAC proteins play an integral role as signaling molecules in the regulation of diverse cellular processes such as cytoskeletal organization, development, membrane trafficking, cell polarity development, morphogenesis, cell differentiation, migration, and response to pathogens. To some extent, diverse functions of Rho family GTPases are attributed to differences in the regulatory mechanisms coordinating upstream and downstream effectors and their contribution in feedback regulation. Recent findings revealed that the developmental responses in plants are mediated by a spatiotemporal coordination between ROP/RAC GTPases and their physical association with functional partner proteins. However, most of the approaches that have been used to understand Rho GTPase signaling do not determine its regulation in dimensions of space and time.

Extending the traditional models, novel imaging techniques have emerged that allow visualization of these cellular signaling events in real time. These have enabled to develop a new insight into the definite functioning of Rho GTPases inside the cells. The new techniques have shown that GTPase mediates interaction within micrometer length scales and within fraction of time. Rho interacting spatiotemporal signaling modules are divided into four classes. First class includes scaffolding proteins: they target Rho GTPases to specific membrane domains. Second class comprises activating GEFs (Guanine nucleotide exchange factors), which catalyze the exchange reaction of GDP with GTP. Third class includes, GAPs (GTPase activating proteins) that stimulate inactivation of GTPases by promoting GTP hydrolysis, while the fourth type of regulators, GDIs (Guanine nucleotide dissociation inhibitors), suppresses nucleotide exchange and mediates sequestration of GTPase in the cytosol from membrane to inhibit its activity. All of these regulatory proteins can perceive upstream signals allowing Rho GTPases to assimilate multiple signals.

© The Author(s) 2015
G.K. Pandey et al., *GTPases*, SpringerBriefs in Plant Science,
DOI 10.1007/978-3-319-11611-2_9

Many studies involving mutants and overexpression have revealed that ROPs have many physiological roles such as directional outgrowth of pollen tubes, cell shaping, root hair elongation, hormone signaling, and innate immunity against pathogen attack [1, 2]. Given that there are multiple functions of ROP proteins in plants, the major challenge is to understand their signaling specificity. The importance of precise spatial localization and timely activation in response to a specific stimulus is emphasized by the fact that its deregulation can cause serious developmental defects. Recent work highlighted the substantial new insights into mechanisms to understand how Rho signaling is regulated within space and time.

Detection of Rho GTPase Activity in Plant Cells

Plant Rho family small GTPases are conserved signaling switches in metazoa and regulate many cellular processes. To understand RhoGTPase functions, it is necessary to determine the fundamental mechanism regulating their activity. Small GTPase activity in plants is controlled internally by plant hormones and on the exterior by environmental stimulus. Instant activation of plant ROPs in response to specific signal suggests a direct regulatory role of these proteins in that particular pathway. Many studies have focused upon developing methods to evaluate the real-time changes in ROP activity in plant cells.

ROP/RAC like GTPase is the only subfamily of RhoGTPases identified in plants [1, 2]. These proteins predominantly function as a signaling switch between GDP-bound inactive and GTP-bound active forms. The RhoGTPases signaling pathway integrates upstream modules through GEFs, GDIs, and GAPs and direct many downstream effectors such as RICs (Rop-interactive CRIB motif containing proteins) and ICRs (interactors of constitutively active ROPs) [3]. The effector protein RICs through their CRIB (Cdc42/Rac-interactive binding) motif specifically interact with activated ROPs and regulate microtubule ordering and cytoskeletal organization in plant cells [4].

The mechanism of ROP/RIC interaction is fundamental for the detection of ROP activity in living cells by several protein–protein interaction assays including fluorescence resonance energy transfer (FRET) analysis and pull-down assays [5–7]. Several approaches have been widely used to dissect the molecular basis of spatiotemporal signaling events with each of them having its own benefits and drawbacks.

FRET Assay

FRET is a powerful microscopic technique to determine protein–protein interactions in living cells. In this assay fluorescence can only be detected when two appropriately chosen fluorophores are less than 10 nm apart from each other. With the advent of instrumentation and multiple colored fluorescent probes this technique has been adapted widely to detect protein interactions in living cells. FRET is equally useful in spatiotemporal elucidation of small GTPase signaling pathway and led to significant novel insights into their physiological role.

FRET involves radiative transfer of energy from the excited dipole of donor to the suitably oriented dipole of the acceptor. The light sensitive fluorophores should be sufficiently close to each other as this technique is extremely sensitive to small changes in distance. The donor fluorophore achieves excited energy state by absorbing a quantum of light, and when the excited electron relaxes to the ground state, the released energy is transferred to the acceptor fluorophore [8, 9]. Moreover, the emission and absorption spectral overlap of the two fluorophores must be coinciding with each other to nullify the cross talk and at the same time for efficient transfer of energy [9–12]. Along with other methods of protein–protein interaction analysis, FRET imaging has emerged as a standard approach to monitor RhoGTPase interaction with effector molecules in living cells with higher specificity and better spatiotemporal resolution. With the advent of GFP-based FRET probes, visualization of the spatiotemporal activities of Rho GTPases in living cells has now turned effortless. FRET-based probes termed as "Ras and interacting protein chimeric unit" (Raichu) probes have been developed with broad applications in number of fields [13]. The structural assembly of Raichu probes consists of four distinct elements including GTPase, GTP-binding domain of binding protein, a donor (CFP), and an acceptor (YFP) linked successively from the N-terminus with the help of spacers [14, 15].

In plants using FRET, the physical interaction between ROP6 and its downstream effector RIC1 (ROP-interactive CRIB motif containing protein (1) was analyzed in living cells. RIC1 is a microtubule associated protein, which promotes ordering of cortical microtubules in the indenting regions of leaf epidermal pavement cells in *Arabidopsis* [16]. The activation of RIC1 for the microtubule organization is regulated by ROP6, another member of Rho GTPase family. RIC1 was fused to CFP (cyan fluorescent protein) and ROP6 was fused to YFP (yellow fluorescent protein). When two proteins were transformed and coexpressed in the *rop6ric1* double mutant cells, strong FRET signals were observed at the cortical region near the plasma membrane confirming their in vivo interaction [17].

Similarly, FRET was used to visualize and quantify interaction between ROP1 and RIC4 in the apical domain of the plasma membrane in the pollen tube. The assay involving CFP-RIC4 and YFP-ROP1 has obtained FRET signals prominently in the tip of pollen tubes suggesting that they primarily interact at the apical region of the plasma membrane [18].

FRET microscopy has also been used to visualize active, GTP-bound Rab5 in living cells. Specialized FRET-based sensors were developed to localize active Rab5 during signaling events. The activated Rab5 fused to CFP and specialized molecular sensors involving Rab5-binding fragments fused to YFP were found to be interacting in endosomal compartments [19]. Thus, the visualization of GTPase spatial activity in vivo with FRET biomolecular biosensors provided the conclusive evidence that the GTPase functions as the key regulator of nucleo-cytoplasmic transport and cytoskeletal arrangement [19–23]. FRET probes and biochemical assay thus provided an insight into the fine-tuning process of the cell that involves precise cross talk between multiple GTPases to specify morphogenic events.

In spite of broad applications and progressive advancements in this method of spatiotemporal detection of proteins, FRET remains to be technically challenging and is generally followed after pull-down assay [16, 17].

Biochemical Assay for the Detection of ROP GTPase Activity

The detection of ROP family GTPase activity is significant when studying signaling events elicited by it. In the past few years, extensive progress has been made in developing biochemical assays, which enables the quantification of average activation level of a given GTPase in a cell. Specialized high affinity probes were developed to monitor the GTPase activity. These probes were developed by fusing glutathione-s-transferase (GST) to high-affinity GTPase binding domain of specific downstream effector proteins. Since the GTPase-binding domain binds preferentially to the active form of ROP GTPases, coupling these domains as probes to beads allows the removal of the active GTP binding complex and subsequent quantification by immunoblotting. Even though the antibody-specific probes for high affinity effector domains do not exist for all small GTPases, analysis of certain members is now regularly performed.

For ROP activity detection by pull-down assay, the high affinity effector domain of RICs fused to maltose-binding protein (MBP) is used to extract GTP bound active RHO GTPase from total protein extracts. Subsequently, western blotting is used to quantify the amount of active ROPs pulled down by RIC effector domains [24, 25]. Since in plants ROP GTPases belong to highly homologous multigene family, a high similarity among each of the member proteins makes it impossible to develop specific antibodies for each of them. To analyze the activity of a specific ROP, it is fused with a GFP tag or marked with a myc epitope. The fused protein is then transformed into plants and isoform specific assay was performed to determine the activity of expressed ROP protein. One of the examples to describe this biochemical assay includes analysis of ROP2 and ROP6 activities in protoplasts in response to auxin. Initially, the GFP-tagged ROP2 and ROP6 were stably expressed in plants. Then, the protoplasts were isolated from leaves of transgenic seedlings and treated with various concentrations of auxin several times and then frozen by liquid nitrogen. The effector protein RIC1 fused with MBP was conjugated to agarose beads and mixed with the protoplast extracts. After incubation for a few hours the beads were washed and the activated GFP-ROP2 or GFP-ROP6 now associated with MBP-RIC1 beads was further processed by western blotting with an anti-GFP antibody. The above analysis thus used to confirm the findings that auxin perception leads to ROP2 and ROP6 activation in the cytoplasmic side of plasma membrane [6]. This approach has clearly established that the ROPs are master regulators for their multifunctional roles in plant cells. However, one of the limitations of this technique is the variability in results due to variation in sample preparation.

Light-Gated Protein Interactions

Some of the recent studies have illustrated a novel approach to study Rho GTPase activity by critically manipulating signaling events in time and space. To provide a solution to the difficulty in manipulating light sensitive proteins with optical reporters, a new approach has been devised that use genetically encoded light-control system centered on an improved phytochrome signaling mechanism in plants. This system uses light-gated translocation of proteins to a subcellular compartment within microscale of space and time [26]. This approach can be potentially used for the designing of various light-controlled system enabling more precise quantitative measurements of signaling activity inside the cells.

Phytochromes are photoreceptive signaling proteins in plants vital for several light sensitive processes in plants such as seed germination. The photoisomerization events involve interaction between phytochromes and downstream transcription factor, phytochrome interaction factor 3 (PIF3). This light sensitive interaction using PhyB-PIF has been modified to construct a genetically encoded reporting system for spatial and temporal control of signaling activity in live cells [26]. In addition, the direct connection between the signaling event and the chosen fluorescent fraction allows quantitative detection of change in activity. Since the system has been used in mammalian cells and originally based on natural mechanism in plants, it is found suitable for most eukaryotic cells.

Thus, this system functions as a novel analytical tool, in which highly complex signaling events can be controlled by high spatial and temporal resolution of light [26]. A key GTPase-regulating actin cytoskeletal dynamics was genetically engineered to develop a photoactivable variant of Rac1 in animals. This was done by fusing photoreactive LOV (light oxygen voltage) domain from phototropin with Rac1. The resulting photoactivable Rac1 could be activated by light to trigger plasma membrane ruffling and protrusion [27]. Considering the endeavor of this approach this could be useful in understanding more sensitive cross talk between signaling modules and as a result will enhance further understanding of Rho GTPase signaling. Furthermore, this will improve further recognition of many Rho GTPase effectors and activators that have not been studied extensively till now expanding the Rho GTPase functional range.

Bacterial Toxins to Study Plant GTPase Signaling

Functional ROP GTPase studies have largely benefited from the experimental modulation of ROP activity in plant cells. The different approaches include generation of overexpression, dominant negative mutants, and antisense RNA ROP constructs in both rice and *Arabidopsis* [28–30]. Recently, an interesting study has put forward a novel concept to study Rac/Rop GTPase signaling using bacterial protein toxins [31]. They have demonstrated that two bacterial toxins CNF1 and toxin B through their catalytic domains cause deamidation and glucosylation of ROPs in vitro.

CNF1 (cytotoxic necrotizing factor 1) is a multidomain protein causing urinary tract infections and is produced by pathogenic *E. coli* strains. CNF1 through its C-terminal glutamine deamidase domain targets animal Rho GTPases for constitutive activation [32–34]. Toxin B from *Clostridium difficile* is a large 270-kDa multidomain protein responsible for causing diarrhea in animals. Toxin B after glucosylating Rho GTPases interferes with their signaling pathway by inhibiting their interaction with the effector molecules [35, 36]. Transient expression of both CNF1 and toxin B in *Arabidopsis* proved to be negatively regulating Rop-mediated leaf morphogenesis and plant growth in general. These results shows that bacterial toxins can be expressed in plant cells and are useful to study Rho signaling pathway.

Significance of Systemic Approaches to Measure Signaling Modularity

Measuring response to specific perturbations and activation of Rho family GTPases has significantly increased our understanding of their function and regulation. The advanced imaging techniques have surpassed the traditional biochemical and cell biological methods to significantly contribute in spatiotemporal regulation of signaling pathways in live cells. The high resolution offered by these new approaches can potentially improve ambiguities in earlier models. Finally, the observation of GTPase activity in endomembrane regions will dissect the stiff regulation mechanism of Rho GTPases with high spatial and temporal resolution providing insight into the critical cell behaviors.

References

1. Yang Z. Small GTPases: versatile signaling switches in plants. Plant Cell. 2002;14(Suppl): S375–88.
2. Gu Y, Wang Z, Yang Z. ROP/RAC GTPase: an old new master regulator for plant signaling. Curr Opin Plant Biol. 2004;7:527–36.
3. Nagawa S, Xu T, Yang Z. RHO GTPase in plants: conservation and invention of regulators and effectors. Small GTPases. 2010;1(2):78–88.
4. Wu G, Gu Y, Li S, Yang Z. A genome-wide analysis of Arabidopsis Rop-interactive CRIB motif-containing proteins that act as Rop GTPase targets. Plant Cell. 2001;13(12):2841–56.
5. Fu Y, Gu Y, Zheng Z, Wasteneys G, Yang Z. Arabidopsis interdigitating cell growth requires two antagonistic pathways with opposing action on cell morphogenesis. Cell. 2005;120(5): 687–700.
6. Xu T, Wen M, Nagawa S, Fu Y, Chen JG, Wu MJ, et al. Cell surface- and rho GTPase-based auxin signaling controls cellular interdigitation in Arabidopsis. Cell. 2010;143(1):99–110.
7. Baxter-Burrell A, Yang Z, Springer PS, Bailey-Serres J. RopGAP4-dependent Rop GTPase rheostat control of Arabidopsis oxygen deprivation tolerance. Science. 2002;296(5575): 2026–8.
8. Wu P, Brand L. Resonance energy transfer: methods and applications. Anal Biochem. 1994;218(1):1–13.

9. Clegg RM. FRET tells us about proximities, distances, orientations and dynamic properties. J Biotechnol. 2002;82:177–9.

10. Truong K, Ikura M. The use of FRET imaging microscopy to detect protein-protein interactions and protein conformational changes in vivo. Curr Opin Struct Biol. 2001;11:573–8.

11. Pollok BA, Heim R. Using GFP in FRET-based applications. Trends Cell Biol. 1999;9:57–60.

12. Kenworthy AK. Imaging protein-protein interactions using fluorescence resonance energy transfer microscopy. Methods. 2001;24(3):289–96.

13. Nakamura T, Kurokawa K, Kiyokawa E, Matsuda M. Analysis of the spatio-temporal activation of Rho GTPases using Raichu probes. Methods Enzymol. 2006;406:315–32.

14. Itoh RE, Kurokawa K, Ohba Y, Yoshizaki H, Mochizuki N, Matsuda M. Activation of rac and cdc42 video imaged by fluorescent resonance energy transfer-based single-molecule probes in the membrane of living cells. Mol Cell Biol. 2002;22(18):6582–91.

15. Yoshizaki H, Ohba Y, Kurokawa K, Itoh RE, Nakamura T, Mochizuki N, et al. Activity of Rho-family GTPases during cell division as visualized with FRET-based probes. J Cell Biol. 2003;162(2):223–32.

16. Fu Y, Gu Y, Zheng Z, Wasteneys G, Yang Z. Arabidopsis interdigitating cell growth requires two antagonistic pathways with opposing action on cell morphogenesis. Cell. 2005;120(5): 687–700.

17. Fu Y, Xu T, Zhu L, Wen M, Yang Z. A ROP GTPase signaling pathway controls cortical microtubule ordering and cell expansion in Arabidopsis. Curr Biol. 2009;19(21):1827–32.

18. Hwang JU, Gu Y, Lee YJ, Yang Z. Oscillatory ROP GTPase activation leads the oscillatory polarized growth of pollen tubes. Mol Biol Cell. 2005;16(11):5385–99.

19. Galperin E, Sorkin A. Visualization of Rab5 activity in living cells by FRET microscopy and influence of plasma-membrane-targeted Rab5 on clathrin-dependent endocytosis. J Cell Sci. 2003;116(Pt 23):4799–810.

20. Kalab P, Weis K, Heald R. Visualization of a Ran-GTP gradient in interphase and mitotic Xenopus egg extracts. Science. 2002;295(5564):2452–6.

21. Kalab P, Pralle A, Isacoff EY, Heald R, Weis K. Analysis of a RanGTP-regulated gradient in mitotic somatic cells. Nature. 2006;440(7084):697–701.

22. Caudron M, Bunt G, Bastiaens P, Karsenti E. Spatial coordination of spindle assembly by chromosome-mediated signaling gradients. Science. 2005;309(5739):1373–6.

23. Dumont J, Petri S, Pellegrin F, Terret ME, Bohnsack MT, Rassinier P, et al. A centriole- and RanGTP-independent spindle assembly pathway in meiosis I of vertebrate oocytes. J Cell Biol. 2007;176(3):295–305.

24. Xu T, Wen M, Nagawa S, Fu Y, Chen JG, Wu MJ, et al. Cell surface- and rho GTPase-based auxin signaling controls cellular interdigitation in Arabidopsis. Cell. 2010;143(1):99–110.

25. Tao LZ, Cheung AY, Wu HM. Plant Rac-like GTPases are activated by auxin and mediate auxin-responsive gene expression. Plant cell. 2002;14(11):2745–60.

26. Levskaya A, Weiner OD, Lim WA, Voigt CA. Spatiotemporal control of cell signalling using a light-switchable protein interaction. Nature. 2009;461:997–1001.

27. Wu YI, Frey D, Lungu OI, Jaehrig A, Schlichting I, Kuhlman B, et al. A genetically encoded photoactivatable Rac controls the motility of living cells. Nature. 2009;461(7260):104–8.

28. Schiene K, Puhler A, Niehaus K. Transgenic tobacco plants that express an antisense construct derived from a Medicago sativa cDNA encoding a Rac-related small GTP-binding protein fail to develop necrotic lesions upon elicitor infiltration. Mol Gen Genet. 2000;263(5):761–70.

29. Miki D, Itoh R, Shimamoto K. RNA silencing of single and multiple members in a gene family of rice. Plant Physiol. 2005;138(4):1903–13.

30. Hoefle C, Huesmann C, Schultheiss H, Bornke F, Hensel G, Kumlehn J, et al. A barley ROP GTPase ACTIVATING PROTEIN associates with microtubules and regulates entry of the barley powdery mildew fungus into leaf epidermal cells. Plant Cell. 2011;23(6):2422–39.

31. Singh MK, Ren F, Giesemann T, Bosco CD, Pasternak TP, Blein T, et al. Modification of plant Rac/Rop GTPase signalling using bacterial toxin transgenes. Plant J. 2013;73(2):314–24.

32. Flatau G, Lemichez E, Gauthier M, Chardin P, Paris S, Fiorentini C, et al. Toxin-induced activation of the G protein p21 Rho by deamidation of glutamine. Nature. 1997;387(6634): 729–33.
33. Schmidt G, Sehr P, Wilm M, Selzer J, Mann M, Aktories K. Gln 63 of Rho is deamidated by Escherichia coli cytotoxic necrotizing factor- 1. Nature. 1997;387:725–9.
34. Aktories K. Bacterial protein toxins that modify host regulatory GTPases. Nat Rev Microbiol. 2011;9:487–98.
35. Just I, Fritz G, Aktories K, Giry M, Popoff MR, Boquet P, et al. Clostridium difficile toxin B acts on the GTP-binding protein Rho. J Biol Chem. 1994;269(14):10706–12.
36. Belyi Y, Aktories K. Bacterial toxin and effector glycosyltransferases. Biochim Biophys Acta. 2010;1800(2):134–43.

Chapter 10
Key Questions and Future Prospects

Introduction

Small GTP-binding proteins are ubiquitous molecular switches that exist in eukaryotes functioning as molecular switches, which cycle between active and inactive states. Plants also contain a large specific class of small GTPases, termed ROP, playing an important role in plant signal transduction mechanism. Plants, on the other hand, lack Ras subfamily proteins altogether with members of other four subfamilies conserved in their genome. In recent years, small GTP-binding proteins have emerged as an intensively studied group of regulators in plants. Rop in plants has been found to regulate an array of physiological processes including pollen tube growth, cytoskeletal arrangement, ROS generation, cell division, response to hormones, and resistance against pathogens. Several evidences have indicated that plants have developed unique molecular mechanisms to control GTPase protein activity predominantly through several upstream regulators and downstream effector proteins.

GTPases and Lipid Interactions

Rho family GTPases are well known to modulate the activity of phospholipases and contrariwise lipids critically regulate them at the functional level. The evolutionary association of lipids and small GTPase-mediated signal transduction events has resulted in an intermingling of these two distinct classes of cellular processes. Consequently, it could be said that generation of phospholipids is vital for Rho GTPase activation and they might be functioning as downstream effectors or needed for signal transduction events. The important aspect of this association that now needs to be addressed is to evaluate the vitality of these interactions under disease and different stress conditions in plants.

© The Author(s) 2015
G.K. Pandey et al., *GTPases*, SpringerBriefs in Plant Science,
DOI 10.1007/978-3-319-11611-2_10

Upstream and Downstream Regulators

Even though numerous functions and actions of the GTPase gene family have been revealed, there are still several questions that need to be answered. The elucidation of mechanism of activation of small GTPases by GEFs and their inactivation by GAPs may shed some light on their role as biotimers rather than as signaling switches. Further, spatiotemporal regulation of the activation and inactivation of these GTPases may indicate their biological regulation. There have been several studies on small GTP-binding proteins in animal. However, Rho GTPases in plants are marginally studied until now. Current progress suggests striking functional similarity between plant and animal GTPases including upstream and downstream effectors to the regulatory networks. Despite this, functional specificity confined only to plants has also been detected in ROP GTPases. The Ras subfamily of GTPases is found only in animals perhaps due to lack of tyrosine kinase receptor in plants.

Future Perspectives

Future work should emphasize on GTPase functional conservation and nonconservation between animals and plants. Some of the key points that need to be addressed include interaction of different members of GTPase subfamilies with one another in the same molecular pathway. Interestingly, some of the reports have suggested the regulation of ROPs by plant hormones. The general perception designates both plant hormones and small GTPases as signaling molecules. The interlinking of the two signaling modules suggests that receptor for plant hormone signaling might be linking the two processes, simultaneously.

Novel GFP-based FRET probes are emerging as promising tools for facilitating better analysis of spatiotemporal dynamics of Rho GTPase activity in living cells. Future work can be directed towards finding their application in physiological systems. Introduction of more sophisticated imaging and detection tools may further enhance the range of applications for these probe-based assays.

The presence of multigene ROP family in plants and equally complex signaling pathways presents a challenge for determining the ROP-dependent signaling activity. The foremost question that remains to be answered is which one out of the numerous signaling pathways in plants is driven by Rop GTPases. Given that multiple ROP proteins are present in plants, does each of them participate in a distinct signaling pathway or mediates cross talk between different signaling pathways? Considering that ROPs act as a switch to oscillate between active and inactive states, the uncertainty remains over the activity regulatory mechanism. Some other important points to be pondered upon include the significance of ROPs in RLK-mediated signaling pathways, downstream targets of ROPs, and how they achieve target specificity.

All these are some of the clues, which require further validation. One gaping hole in our knowledge of plant ROPs is their functional characterization. The characterization of GTPase superfamily in model plants like *Arabidopsis* using genetic, genomic, bioinformatic, and biochemical methods needs to be performed to yield information regarding aspects of functional diversity and signaling network. With the advent of various spatiotemporal techniques capable of detecting signaling activity in vivo conditions, finding answers to these questions may soon turn into reality.

Small GTPases in both animal and plants are involved critically in the regulation of polar growth such as axon growth in animals and pollen tube and root hair cell elongation in plants. Despite several mechanistic differences, quite a few common key factors or regulators exist such as small GTPase complement. Due to ease of work with plants system (robust genetic tools and ease of genetically manipulating plants compared to animals), it could be possible to identify the holistic regulation of the polar cell growth in plant system, and the knowledge could be useful to understand and cure several diseases and neurodegenerative disorder where polar cell growth is affected in animals.